ents

Unit 1: Pure Mathematics

CHAPTER 1: SIMPLIFYING ALGEBRAIC EXPRESSIONS

1.1 Adding And Subtracting Algebraic Expressions

When adding or subtracting you need to find the lowest common denominator. For example, simplify:

1. $\dfrac{2x + 3}{5} + \dfrac{2x - 3}{6}$

The lowest common denominator of 5 and 6 is 30
So we change each fraction to a denominator of 30

This gives: $\dfrac{12x + 18}{30} + \dfrac{10x - 15}{30}$

We can then add these fractions to get: $\dfrac{22x + 3}{30}$

....

2. $\dfrac{2x - 5}{6} - \dfrac{4 - 2x}{3}$

The lowest common denominator of 6 and 3 is 6
So we change the second fraction to a denominator of 6

This gives: $\dfrac{2x - 5}{6} - \dfrac{8 - 4x}{6}$

We can then subtract these fractions to get: $\dfrac{6x - 13}{6}$

....

3. $\dfrac{3}{5x - 2} + \dfrac{2}{1 - 2x}$

The lowest common denominator is $(5x - 2)(1 - 2x)$
So we change each fraction to a denominator of $(5x - 2)(1 - 2x)$

This gives: $\dfrac{3 - 6x}{(5x - 2)(1 - 2x)} + \dfrac{10x - 4}{(5x - 2)(1 - 2x)}$

We can then add these fractions to get: $\dfrac{4x - 1}{(5x - 2)(1 - 2x)}$

You do not need to work out the brackets on the denominator.

1.2 Adding And Subtracting Algebraic Fractions Where Each Denominator Has A Product Of 2 Terms

For example, simplify: **1.** $\dfrac{6}{(x + 5)(2x - 3)} + \dfrac{2}{x(2x - 3)}$

The lowest common denominator is $x(x + 5)(2x - 3)$
So we change each fraction to a denominator of $x(x + 5)(2x - 3)$

This gives: $\dfrac{6x}{x(x + 5)(2x - 3)} + \dfrac{2x + 10}{x(x + 5)(2x - 3)}$

We can then add these fractions to get: $\dfrac{8x + 10}{x(x + 5)(2x - 3)}$

You do not need to work out the brackets on the denominator.

....

2. $\dfrac{5}{(x - 6)(3x + 2)} - \dfrac{3}{(x + 4)(x - 6)}$

The lowest common denominator is $(x + 4)(x - 6)(3x + 2)$
So we change each fraction to a denominator of $(x + 4)(x - 6)(3x + 2)$

This gives:

$$\frac{5x + 20}{(x + 4)(x - 6)(3x + 2)} - \frac{9x + 6}{(x + 4)(x - 6)(3x + 2)}$$

We can then subtract these fractions to get:

$$\frac{-4x + 14}{(x + 4)(x - 6)(3x + 2)}$$

Exercise 1B: *Simplify the following*

1. $\dfrac{7}{(x + 2)(x - 3)} + \dfrac{4}{x(x - 3)}$

2. $\dfrac{5}{2x(x - 3)} + \dfrac{2}{(x + 2)(x - 3)}$

3. $\dfrac{4}{(x + 5)(x + 3)} + \dfrac{3}{(x - 1)(x + 5)}$

4. $\dfrac{5}{(x - 2)(2x + 1)} + \dfrac{1}{(x - 2)(x + 4)}$

5. $\dfrac{3}{x(x - 4)} + \dfrac{2}{x(x + 2)}$

6. $\dfrac{4}{(x - 2)(x + 1)} - \dfrac{3}{(x + 1)(x + 5)}$

7. $\dfrac{5}{2x(x - 3)} - \dfrac{2}{(x + 4)(x - 3)}$

8. $\dfrac{3}{(x - 1)(x + 4)} - \dfrac{4}{(x + 4)(x - 2)}$

9. $\dfrac{2}{(x + 5)(x - 2)} - \dfrac{3}{(x + 5)(x + 2)}$

10. $\dfrac{3}{(x + 2)(2x - 1)} - \dfrac{2}{(x - 4)(2x - 1)}$

1.3 Adding And Subtracting Algebraic Fractions With Quadratic Denominators

Factorise the denominators. Then add or subtract as before. For example, simplify:

1. $\dfrac{4}{x^2 - 3x} + \dfrac{2}{x^2 - x - 6}$

but $x^2 - 3x = x(x - 3)$ by common factors and $x^2 - x - 6 = (x + 2)(x - 3)$ by quadratic factors

So we can write the expression as

$$\frac{4}{x(x - 3)} + \frac{2}{(x + 2)(x - 3)}$$

The lowest common denominator is $x(x + 2)(x - 3)$ So we change each fraction to a denominator of $x(x + 2)(x - 3)$

This gives: $\dfrac{4x + 8}{x(x + 2)(x - 3)} + \dfrac{2x}{x(x + 2)(x - 3)}$

We can then add these fractions to get: $\dfrac{6x + 8}{x(x + 2)(x - 3)}$

....

2. $\dfrac{5}{3x^2 + 10x - 8} - \dfrac{2}{x^2 - 16}$

but $3x^2 + 10x - 8 = (3x - 2)(x + 4)$ by quadratic factors and $x^2 - 16 = (x - 4)(x + 4)$ by difference of two squares

So we can write the expression as:

$$\frac{5}{(3x - 2)(x + 4)} - \frac{2}{(x - 4)(x + 4)}$$

The lowest common denominator is $(x - 4)(x + 4)(3x - 2)$ So we change each fraction to a denominator of $(x - 4)(x + 4)(3x - 2)$

This gives:

$$\frac{5x - 20}{(x - 4)(x + 4)(3x - 2)} - \frac{6x - 4}{(x - 4)(x + 4)(3x - 2)}$$

We can then subtract these fractions to get:

$$\frac{-x - 16}{(x - 4)(x + 4)(3x - 2)}$$

Exercise 1C: *Simplify the following*

1. $\dfrac{4}{3x^2 + 6x} + \dfrac{3}{x^2 + 3x + 2}$

2. $\dfrac{5}{x^2 + 2x - 8} + \dfrac{2}{x^2 - 4}$

3. $\dfrac{3}{2x^2 + 7x - 4} + \dfrac{5}{x^2 + x - 12}$

4. $\dfrac{3}{2x^2 - 10x} + \dfrac{2}{2x^2 - 11x + 5}$

5. $\dfrac{4}{x^2 + 3x - 10} + \dfrac{3}{x^2 - 7x + 10}$

6. $\dfrac{4}{2x^2 - x} - \dfrac{2}{x^2 + x}$

7. $\dfrac{3}{x^2 - 2x - 8} - \dfrac{2}{x^2 - 16}$

8. $\dfrac{2}{2x^2 - x - 6} - \dfrac{3}{x^2 - 2x}$

9. $\dfrac{4}{x^2 + x - 12} - \dfrac{5}{2x^2 + 13x + 20}$

10. $\dfrac{3}{4x^2 + 7x - 2} - \dfrac{2}{x^2 - x - 6}$

1.4 Multiplying And Dividing Algebraic Expressions

Multiplying
Cancel as far as possible.
Then multiply the numerators and then multiply the denominators.

For example: $\dfrac{6x^3}{y^4t^2} \times \dfrac{10y^2}{8tx}$

You can divide 2 into 6 and 8 to get 3 and 4
You can then divide 2 into 10 and 4 to get 5 and 2
You can cancel x^3 and x by x to get to x^2 and 1
You can cancel y^2 and y^4 by y^2 to get to 1 and y^2

This gives $\dfrac{3x^2}{y^2t^2} \times \dfrac{5}{2t}$

You can then multiply these to get the answer $\dfrac{15x^2}{2y^2t^3}$

Dividing
Dividing by a fraction is the same as multiplying by its reciprocal.

For example: $\dfrac{10a^3b}{6b^2} \div \dfrac{4c}{5a^2}$

The reciprocal of $\dfrac{4c}{5a^2}$ is $\dfrac{5a^2}{4c}$

So we can multiply $\dfrac{10a^3b}{6b^2} \times \dfrac{5a^2}{4c}$

You can divide 2 into 10 and 4 to get 5 and 2
You can cancel b and b^2 by b to get to 1 and b

This gives $\dfrac{5a^3}{6b} \times \dfrac{5a^2}{2c}$

You can then multiply these to get the answer $\dfrac{25a^5}{12bc}$

Exercise 1D: *Simplify the following*

1. $\dfrac{5x^2}{yv} \times \dfrac{2y^2}{vx}$

2. $\dfrac{3vt}{2n^2} \times \dfrac{4n}{6v^2}$

3. $\dfrac{10q}{r^2} \times \dfrac{2qr}{5n}$

4. $\dfrac{4vw}{2t} \times \dfrac{9t^2}{2v^3}$

5. $\dfrac{6qr}{2v^2} \times \dfrac{10vw}{3q^2}$

6. $\dfrac{9a^2}{5b} \div \dfrac{3c}{2ab^3}$

7. $\dfrac{4q^2}{3r} \div \dfrac{5q}{6rn}$

8. $\dfrac{3t^2}{2vw} \div \dfrac{9tn}{4v^2}$

9. $\dfrac{9a}{2b^2} \div \dfrac{7a^2}{4bc}$

10. $\dfrac{4v^2}{3q} \div \dfrac{2r}{5vq^2}$

1.5 Dividing Algebraic Fractions With Quadratic Numerators And Denominators

Factorise each quadratic expression. Then cancel as far as possible. For example, simplify:

1. $\dfrac{14x^2 - 7x}{4x^2 - 1}$

$14x^2 - 7x = 7x(2x - 1)$ by common factors and

$4x^2 - 1 = (2x - 1)(2x + 1)$ by difference of two squares

So you can rewrite the fraction as $\dfrac{7x(2x - 1)}{(2x - 1)(2x + 1)}$

You can then cancel out the $(2x - 1)$
to get the final answer:

$$\dfrac{7x}{2x + 1}$$

2. $\dfrac{x^2 + 2x - 15}{3x^2 + 13x - 10}$

$x^2 + 2x - 15 = (x - 3)(x + 5)$ by quadratic factors and
$3x^2 + 13x - 10 = (3x - 2)(x + 5)$ by quadratic factors

So you can rewrite the fraction as $\dfrac{(x - 3)(x + 5)}{(3x - 2)(x + 5)}$

You can then cancel out the $(x + 5)$ to get the final answer

$$\dfrac{x - 3}{3x - 2}$$

Exercise 1E: Simplify the following

1. $\dfrac{2x^2 + 5x - 3}{x^2 + x - 6}$

2. $\dfrac{3x^2 - 6x}{2x^2 - 5x + 2}$

3. $\dfrac{x^2 - 25}{x^2 - x - 20}$

4. $\dfrac{4x^2 + 5x - 6}{x^2 - x - 6}$

5. $\dfrac{2x^2 + 9x - 5}{2x^2 + 3x - 2}$

6. $\dfrac{4x^2 - 12x}{x^2 - x - 6}$

7. $\dfrac{2x^2 + 7x - 4}{x^2 + 2x - 8}$

8. $\dfrac{3x^2 + 13x - 10}{6x^2 + 5x - 6}$

9. $\dfrac{8x^2 - 4x}{4x^2 - 1}$

10. $\dfrac{5x^2 + 13x - 6}{x^2 - 2x - 15}$

1.6 More Multiplying And Dividing

For example, work out:

1. $\dfrac{2q^2 - 7q + 3}{4} \times \dfrac{6}{q^2 - 9}$

You need to factorise each quadratic expression:
$2q^2 - 7q + 3 = (q - 3)(2q - 1)$ by quadratic factors and
$q^2 - 9 = (q - 3)(q + 3)$ by difference of two squares
So you can rewrite the multiplication as:
$$\dfrac{(q - 3)(2q - 1)}{4} \times \dfrac{6}{(q - 3)(q + 3)}$$

You can divide 6 and 4 by 2 to get 3 and 2
You can then cancel out the $(q - 3)$ to get the final answer
$$\dfrac{3(2q - 1)}{2(q + 3)}$$

2. $\dfrac{x^2 + 4x}{5} \div \dfrac{x^2 + 2x - 8}{10}$

The reciprocal of $\dfrac{x^2 + 2x - 8}{10}$ is $\dfrac{10}{x^2 + 2x - 8}$

So we can multiply. $\dfrac{x^2 + 4x}{5} \times \dfrac{10}{x^2 + 2x - 8}$

You need to factorise each quadratic expression
$x^2 + 4x = x(x + 4)$ by common factors and
$x^2 + 2x - 8 = (x + 4)(x - 2)$ by quadratic factors
So you can rewrite the multiplication as

$$\dfrac{x(x + 4)}{5} \times \dfrac{10}{(x + 4)(x - 2)}$$

You can divide 10 and 5 by 5 to get 2 and 1
You can then cancel out the $(x + 4)$ to get the final answer

$$\dfrac{2x}{x - 2}$$

Exercise 1F: Work out the following

1. $\dfrac{x^2 - 2x}{3} \times \dfrac{6}{x^2 + 2x - 8}$

2. $\dfrac{2x^2 - 5x - 3}{4} \times \dfrac{10}{6x^2 - x - 2}$

3. $\dfrac{x^2 + 2x - 15}{8} \times \dfrac{6}{x^2 - 5x + 6}$

4. $\dfrac{3x^2 + 14x + 8}{6} \times \dfrac{9}{x^2 + 2x - 8}$

5. $\dfrac{4x^2 - 10x}{10} \times \dfrac{8}{3x^2 + 12x}$

6. $\dfrac{4x^2 + 8x}{3} \div \dfrac{3x^2 - x}{6}$

7. $\dfrac{x^2 - 9x + 14}{4} \div \dfrac{x^2 + 3x - 10}{2}$

8. $\dfrac{2x^2 + 9x - 5}{8} \div \dfrac{8x^2 + 2x - 3}{6}$

9. $\dfrac{x^2 + 9x + 18}{9} \div \dfrac{x^2 + x - 6}{12}$

10. $\dfrac{4x^2 + 7x - 2}{4} \div \dfrac{x^2 - 4}{6}$

CHAPTER 2: EQUATIONS

2.1 Quadratic Equations

A quadratic equation is an equation where the highest power of the variable is 2. You can solve quadratic equations by factorising **but only when it can be factorised.** You can also solve quadratic equations by using the quadratic formula.

Factorising

For example, solve the following:

1. $4x^2 - 25 = 0$

You can factorise $4x^2 - 25$ by difference of 2 squares to get: $(2x - 5)(2x + 5) = 0$

So either 　$2x - 5 = 0$

　　　　　　$2x = 5$

　　　　　　$x = 2\frac{1}{2}$ or 2.5

or 　　　　$2x + 5 = 0$

　　　　　　$2x = -5$

　　　　　　$x = -2\frac{1}{2}$ or -2.5

....

2. $4x^2 - 7x = 0$

You can factorise $4x^2 - 7x$ by common factors to get: $x(4x - 7) = 0$

So either 　$x = 0$

or 　　　　$4x - 7 = 0$

　　　　　　$4x = 7$

　　　　　　$x = 1\frac{3}{4}$ or 1.75

....

3. $2x^2 + 9x - 5 = 0$

You can factorise $2x^2 + 9x - 5$ by quadratic factors to get: $(2x - 1)(x + 5) = 0$

So either 　$2x - 1 = 0$

　　　　　　$2x = 1$

　　　　　　$x = \frac{1}{2}$ or 0.5

or 　　　　$x + 5 = 0$

　　　　　　$x = -5$

Exercise 2A: *Solve the following by factorising*

1. $9x^2 - 16 = 0$ 　　　　**7.** $4x^2 + 17x - 15 = 0$

2. $3x^2 - x - 2 = 0$ 　　**8.** $6x^2 - 14x = 0$

3. $12x^2 + 42x = 0$ 　　**9.** $6x^2 + 23x - 18 = 0$

4. $2x^2 - 7x - 15 = 0$ 　**10.** $2x^2 - 17x + 21 = 0$

5. $6x^2 - x - 2 = 0$ 　　**11.** $6x^2 + 7x + 2 = 0$

6. $3x^2 - 10x - 8 = 0$ 　**12.** $25x^2 - 4 = 0$

Using The Quadratic Formula

The solutions to $ax^2 + bx + c = 0$ are found from substituting the values of a, b and c into the formula

$$\frac{-b \pm \sqrt{b^2 - 4ac}}{2a}$$

For example, solve the following, giving your answers correct to 2 decimal places:

1. $2x^2 + x - 11 = 0$

　　　　　　$a = 2$

　　　　　　$b = 1$

　　　　　　$c = -11$

So 　　　$x = \dfrac{-1 \pm \sqrt{1^2 - 4 \times 2 \times (-11)}}{2 \times 2}$

　　　　　　$x = \dfrac{-1 \pm \sqrt{89}}{4}$

　　　　　　$x = \dfrac{-1 + \sqrt{89}}{4}$ or $\dfrac{-1 - \sqrt{89}}{4}$

　　　　　　$x = 2.108$ or -2.608

So $x = 2.11$ or -2.61, to 2 decimal places.

....

2. $3x^2 - 6x + 1 = 0$

　　　　　　$a = 3$

　　　　　　$b = -6$

　　　　　　$c = 1$

So 　　　$x = \dfrac{6 \pm \sqrt{36 - 4 \times 3 \times 1}}{2 \times 3}$

　　　　　　$x = \dfrac{6 \pm \sqrt{36 - 12}}{6}$

　　　　　　$x = \dfrac{6 \pm \sqrt{24}}{6}$

　　　　　　$x = \dfrac{6 + \sqrt{24}}{6}$ or $\dfrac{6 - \sqrt{24}}{6}$

　　　　　　$x = 1.816$ or 0.184

So $x = 1.82$ or 0.18, to 2 decimal places.

1. $x^2 - 6x + 3 = 0$

2. $2x^2 + x - 7 = 0$

3. $3x^2 - 2x - 9 = 0$

4. $x^2 + 9x + 2 = 0$

5. $2x^2 - x - 7 = 0$

6. $3x^2 + 2x - 2 = 0$

7. $5x^2 - x - 8 = 0$

8. $2x^2 + 8x + 2 = 0$

9. $4x^2 + 5x - 3 = 0$

10. $x^2 - 2x - 9 = 0$

2.2 Fractional Equations

For example, solve:

1. $\dfrac{6}{x-3} + \dfrac{2}{x-4} = 5$

You need to add the fractions by using the lowest common denominator, which is $(x-3)(x-4)$

So we get $\dfrac{6(x-4) + 2(x-3)}{(x-3)(x-4)} = 5$

You now need to work out and simplify the numerator
$6(x-4) + 2(x-3) = 6x - 24 + 2x - 6 = 8x - 30$

This gives $\dfrac{8x - 30}{(x-3)(x-4)} = 5$

You now need to work out and simplify the denominator
$(x-3)(x-4) = x^2 - 7x + 12$

This gives $\dfrac{8x - 30}{x^2 - 7x + 12} = 5$

You can now cross multiply to get
$8x - 30 = 5(x^2 - 7x + 12)$

You can now work out the brackets and then put all the terms on one side (with the x^2 term first and positive because it is a quadratic equation).

$$8x - 30 = 5x^2 - 35x + 60$$
$$0 = 5x^2 - 35x - 8x + 60 + 30$$
So $5x^2 - 43x + 90 = 0$

You can now solve this equation by factorising or using the formula to get
$$(x-5)(5x - 18) = 0$$

So either $\quad x - 5 = 0 \quad\quad x = 5$

or $\quad\quad 5x - 18 = 0 \quad 5x = 18 \quad x = \dfrac{18}{5}$

2. $\dfrac{4}{x-1} - \dfrac{2}{1-x} = 3$

You need to add the fractions by using the lowest common denominator which is $(x-1)(1-x)$

So we get $\dfrac{4(1-x) - 2(x-1)}{(x-1)(1-x)} = 3$

You now need to work out and simplify the numerator
$4(1-x) - 2(x-1) = 4 - 4x - 2x + 2 = -6x + 6$

This gives $\dfrac{-6x + 6}{(x-1)(1-x)} = 3$

You now need to work out and simplify the denominator
$(x-1)(1-x) = -x^2 + 2x - 1$

This gives $\dfrac{-6x + 6}{-x^2 + 2x - 1} = 3$

You can now cross multiply to get
$$-6x + 6 = 3(-x^2 + 2x - 1)$$

You can now work out the brackets and then put all the terms on one side (with the x^2 term first and positive because it is a quadratic equation).

$$-6x + 6 = -3x^2 + 6x - 3$$
$$3x^2 - 12x + 9 = 0$$

You can divide all these terms by 3 to get

$$x^2 - 4x + 3 = 0$$

You can now solve this equation by factorising or using the formula to get

$$(x-3)(x-1) = 0$$
So either $\quad x - 3 = 0$
$$x = 3$$
or $\quad\quad x - 1 = 0$
$$x = 1$$

1. $\dfrac{4}{x-2} + \dfrac{3}{2x-5} = 3$

2. $\dfrac{6}{2x-1} - \dfrac{4}{1-x} = 6$

3. $\dfrac{4}{3x-2} - \dfrac{6}{x+1} = 1$

4. $\dfrac{2}{x-2} + \dfrac{3}{2x-3} = 3$

5. $\dfrac{3}{x+1} - \dfrac{6}{1-2x} = 3$

6. $\dfrac{4}{x-3} - \dfrac{4}{1-x} = 3$

7. $\dfrac{6}{x-1} - \dfrac{5}{1-2x} = 4$

8. $\dfrac{3}{x+1} + \dfrac{6}{2x-1} = 3$

9. $\dfrac{4}{x-1} - \dfrac{2}{2-x} = 4$

10. $\dfrac{3}{x-2} - \dfrac{2}{3-x} = 2$

2.3 Fractional Equations With Both Numerator And Denominator Expressions Involving x

For example, solve:

1. $\dfrac{x+1}{x-1} + \dfrac{3x-1}{2x+1} = 4$

You need to add the fractions by using the lowest common denominator which is $(x-1)(2x+1)$

So we get $\dfrac{(x+1)(2x+1) + (3x-1)(x-1)}{(x-1)(2x+1)} = 4$

You now need to work out and simplify the numerator

$(x+1)(2x+1) = 2x^2 + 3x + 1$ and
$(3x-1)(x-1) = 3x^2 - 4x + 1$

So the numerator is
$2x^2 + 3x + 1 + 3x^2 - 4x + 1 = 5x^2 - x + 2$

This gives $\dfrac{5x^2 - x + 2}{(x-1)(2x+1)} = 4$

You now need to work out and simplify the denominator
$(x-1)(2x+1) = 2x^2 - x - 1$

This gives: $\dfrac{5x^2 - x + 2}{2x^2 - x - 1} = 4$

You can now cross multiply to get
$5x^2 - x + 2 = 4(2x^2 - x - 1)$

You can now work out the brackets and then put all the terms on one side (with the x^2 term first and positive, as it is a quadratic equation)

$$5x^2 - x + 2 = 8x^2 - 4x - 4$$
$$0 = 8x^2 - 5x^2 - 4x + x - 4 - 2$$
So $\quad 3x^2 - 3x - 6 = 0$

You can divide all these terms by 3 to get:
$$x^2 - x - 2 = 0$$
You can now solve this equation by factorising or using the formula to get:
$$(x-2)(x+1) = 0$$

So either $\quad x - 2 = 0$
$$x = 2$$
or $\quad x + 1 = 0$
$$x = -1$$

......

2. $\dfrac{x+3}{x-1} - \dfrac{x-2}{1-2x} = 5$

You need to add the fractions by using the lowest common denominator which is $(x-1)(1-2x)$

So we get $\dfrac{(x+3)(1-2x) - (x-2)(x-1)}{(x-1)(1-2x)} = 5$

You now need to work out and simplify the numerator
$(x+3)(1-2x) = -2x^2 - 5x + 3$ and
$(x-2)(x-1) = x^2 - 3x + 2$

So the numerator is
$-2x^2 - 5x + 3 - (x^2 - 3x + 2)$
$= -2x^2 - 5x + 3 - x^2 + 3x - 2$
$= -3x^2 - 2x + 1$

This gives $\dfrac{-3x^2 - 2x + 1}{(x-1)(1-2x)} = 5$

You now need to work out and simplify the denominator
$(x-1)(1-2x) = -2x^2 + 3x - 1$

This gives $\dfrac{-3x^2 - 2x + 1}{-2x^2 + 3x - 1} = 5$

You can now cross multiply to get:
$-3x^2 - 2x + 1 = 5(-2x^2 + 3x - 1)$

You can now work out the brackets and then put all the terms on one side (with the x^2 term first and positive, as it is a quadratic equation)

$$-3x^2 - 2x + 1 = -10x^2 + 15x - 5$$
$$10x^2 - 3x^2 - 2x - 15x + 1 + 5 = 0$$

So $\qquad 7x^2 - 17x + 6 = 0$

You can now solve this equation by factorising or using the formula to get $(x - 2)(7x - 3) = 0$

So either $\qquad x - 2 = 0$
$$x = 2$$

or $\qquad 7x - 3 = 0$
$$7x = 3$$
$$x = \frac{3}{7}$$

Exercise 2D: *Solve the following:*

1. $\dfrac{x + 1}{x - 2} + \dfrac{3x - 1}{x + 1} = 6$

2. $\dfrac{x + 7}{1 - x} + \dfrac{2}{x + 2} = -1$

3. $\dfrac{x + 3}{3 - x} - \dfrac{1 - 2x}{x - 2} = -1$

4. $\dfrac{x + 4}{x - 2} - \dfrac{2x - 5}{x - 1} = 3$

5. $\dfrac{4 - x}{x + 4} + \dfrac{x + 2}{x + 1} = 3$

6. $\dfrac{x + 3}{2x - 3} - \dfrac{3x - 2}{x - 1} = 2$

7. $\dfrac{x + 1}{3x - 1} + \dfrac{4x - 2}{x + 1} = 3$

8. $\dfrac{x + 1}{x + 3} - \dfrac{3x + 2}{1 - x} = 5$

9. $\dfrac{2x}{x - 2} + \dfrac{1 - 3x}{x - 1} = 2$

10. $\dfrac{x + 3}{x - 3} - \dfrac{x + 4}{4 - x} = 8$

2.4 Completing The Square

Another way to solve a quadratic equation is to complete the square for the variable as shown below.

For example, solve the following by completing the square:

1. $x^2 + 8x - 5 = 0$

To complete the square we must take half of the coefficient of the x term as follows:

½ of 8 = 4 and expand $(x + 4)^2$

Now $(x + 4)^2 = x^2 + 4x + 4x + 16 = x^2 + 8x + 16$

So we can replace $x^2 + 8x$ with $(x + 4)^2 - 16$

This gives
$$(x + 4)^2 - 16 - 5 = 0$$
$$(x + 4)^2 = 16 + 5 = 21$$

So $\qquad x + 4 = \pm\sqrt{21}$
$$x = -4 \pm\sqrt{21}$$

The general rule for solving $x^2 + bx + c = 0$ by completing the square is:

a. Work out ½b

b. Rewrite $x^2 + bx + c = 0$ as $(x + ½b)^2 - (½b)^2 + c = 0$

....

2. $x^2 - 9x + 2 = 0$

To complete the square we must take half of the coefficient of the x term as follows:

½ of $-9 = \dfrac{-9}{2}$ and expand $\left(x - \dfrac{9}{2}\right)^2$

Now $\left(x - \dfrac{9}{2}\right)^2 = x^2 - \dfrac{9}{2}x - \dfrac{9}{2}x + \left(\dfrac{9}{2}\right)^2 = x^2 - 9x + \dfrac{81}{4}$

So we can replace $x^2 - 9x$ with $\left(x - \dfrac{9}{2}\right)^2 - \dfrac{81}{4}$

This gives
$$\left(x - \dfrac{9}{2}\right)^2 - \dfrac{81}{4} + 2 = 0$$

$$\left(x - \dfrac{9}{2}\right)^2 = \dfrac{81}{4} - 2 = \dfrac{81}{4} - \dfrac{8}{4} = \dfrac{73}{4}$$

So $\qquad x - \dfrac{9}{2} = \pm\sqrt{\dfrac{73}{4}} = \dfrac{\pm\sqrt{73}}{2}$

$$x = \dfrac{9}{2} \pm \dfrac{\sqrt{73}}{2}$$

So $\qquad x = \dfrac{9 \pm \sqrt{73}}{2}$

Exercise 2E: *Solve the following by completing the square:*

1. $x^2 + 6x - 3 = 0$
2. $x^2 - 2x - 5 = 0$
3. $x^2 + x - 4 = 0$
4. $x^2 - 3x - 5 = 0$
5. $x^2 + 8x + 5 = 0$
6. $x^2 - 10x - 2 = 0$
7. $x^2 + 5x - 4 = 0$
8. $x^2 - 7x - 5 = 0$
9. $x^2 + 4x - 9 = 0$
10. $x^2 - 10x - 3 = 0$

2.5 Finding Minimum Points By Completing The Square

You can find minimum turning points and the value of x for which they occur by completing the square. For example:

1. Find (i) the minimum value of $y = x^2 - 8x - 5$ by completing the square, and (ii) the value of x at this minimum value.

(i) $y = x^2 - 8x - 5$
So $y = (x - 4)^2 - 16 - 5$
So $y = (x - 4)^2 - 21$

The smallest possible value of $(x - 4)^2$ is 0 (as all square numbers are positive except 0). So the minimum value of $y = x^2 - 8x - 5$ is $0 - 21 = -21$

(ii) $(x - 4)^2 = 0$ when $x - 4 = 0$
     i.e. when $x = 4$

Exercise 2F:

Find (i) the minimum value of the following expressions by completing the square,
(ii) the value of x at each minimum value.

1. $y = x^2 + 6x - 3$
2. $y = x^2 - 4x + 1$
3. $y = x^2 + 8x + 5$
4. $y = x^2 - 2x - 8$
5. $y = x^2 - x - 3$
6. $y = x^2 + 3x + 8$
7. $y = x^2 - 6x + 11$
8. $y = x^2 + 10x + 35$

CHAPTER 3: SIMULTANEOUS EQUATIONS

3.1 Simultaneous Equations With Three Unknowns

These can be solved by using these rules:
a. Eliminate one unknown from 2 of these equations.
b. Eliminate the same unknown from another 2 of these equations.
c. This leaves you with 2 equations with 2 unknowns.
For example, solve the following:

1.
$$3x + 2y - z = 64 \quad (1)$$
$$2x + y + 3z = 84 \quad (2)$$
$$4x - 3y + 2z = 80 \quad (3)$$

Let us get rid of z
You can multiply equation (1) by 3 and then add this to equation (2):

$$9x + 6y - 3z = 192 \quad (4)$$
$$+ \quad 2x + y + 3z = \quad 84 \quad (2)$$
$$\overline{11x + 7y \quad\quad\quad = 276 \quad (5)}$$

You can multiply equation (1) by 2 and then add this to equation (3):

$$6x + 4y - 2z = 128 \quad (6)$$
$$+ \quad 4x - 3y + 2z = \quad 80 \quad (3)$$
$$\overline{10x + y \quad\quad\quad = 208 \quad (7)}$$

You now have to solve the following equations:
$$11x + 7y = 276 \quad (5)$$
$$10x + y = 208 \quad (7)$$

You can multiply equation (7) by 7 and then subtract equation (5):

$$70x + 7y = 1456 \quad (8)$$
$$- \quad 11x + 7y = \quad 276 \quad (5)$$
$$\overline{59x \quad\quad = 1180}$$
$$\text{So } x = \frac{1180}{50} = 20$$

Substituting $x = 20$ into (7) gives
$$10x + y = 208$$
$$200 + y = 208$$
$$y = 208 - 200$$
$$y = 8$$

Substituting $x = 20$ and $y = 8$ into (2) gives
$$2x + y + 3z = 84$$
$$40 + 8 + 3z = 84$$
$$3z = 84 - 40 - 8$$
$$3z = 36$$
$$z = \frac{36}{3} = 12$$

You can check your answers by substituting $x = 20$, $y = 8$ and $z = 12$ into (1):
$$3x + 2y - z = 64$$
$$60 + 16 - 12 = 64$$

2.
$$4x + 2y + 3z = 58 \quad (1)$$
$$3x + 4y + 5z = 70 \quad (2)$$
$$2x - y + 4z = 36 \quad (3)$$

Let us get rid of y
You can multiply equation (3) by 2 and then add this to equation (1):
$$4x - 2y + 8z = 72 \ (4)$$
$$+ \ \underline{4x + 2y + 3z = 58} \ (1)$$
$$8x + \quad 11z = 130 \ (5)$$

You can multiply equation (1) by 2 and then subtract equation (2):
$$8x + 4y + 6z = 116 \quad (6)$$
$$- \ \underline{3x + 4y + 5z = 70} \quad (2)$$
$$5x + \qquad z = 46 \quad (7)$$

You now have to solve the following equations:
$$8x + 11z = 130 \quad (5)$$
$$5x + z = 46 \quad (7)$$

You can multiply equation (7) by 11 and then subtract equation (5):
$$55x + 11z = 506 \quad (8)$$
$$- \ \underline{8x + 11z = 130} \quad (5)$$
$$47x \qquad = 376$$
So $\quad x = \frac{376}{47} = 8$

Substituting $x = 8$ into (7) gives
$$5x + z = 46$$
$$40 + z = 46$$
$$z = 46 - 40$$
$$z = 6$$

Substituting $x = 8$ and $z = 6$ into (1) gives
$$4x + 2y + 3z = 58$$
$$32 + 2y + 18 = 58$$
$$2y = 58 - 32 - 18$$
$$2y = 8$$
$$y = \frac{8}{2} = 4$$

You can check your answers by substituting $x = 8$, $y = 4$ and $z = 6$ into (2):
$$3x + 4y + 5z = 70$$
$$24 + 16 + 30 = 70$$

Exercise 3A: *Solve the following:*

1.	$3x + 2y + 4z = 33$ $x - 3y + 2z = 10$ $2x + y - 3z = -1$	**6.**	$2x - y + 3z = 5$ $3x + 2y - z = -4$ $x - 3y - 2z = -19$
2.	$3x + y - 4z = -9$ $x - 2y + z = 4$ $2x + 3y - 2z = 2$	**7.**	$2x - 3y + z = 8$ $3x + y - 2z = 9$ $x + 2y - 5z = -11$
3.	$4x - y + 2z = 20$ $2x - 3y - z = -1$ $x + 2y - 3z = -1$	**8.**	$4x + 3y - z = 14$ $x + 2y + 3z = 16$ $2x - y + z = 22$
4.	$3x + y - 2z = -5$ $x - 2y - 5z = -3$ $2x + 3y + z = -8$	**9.**	$x - 3y + z = -15$ $2x + y - 3z = 23$ $3x - 2y - z = 3$
5.	$2x - 3y + z = 9$ $x + y - 2z = 7$ $3x - 2y + 4z = 6$	**10.**	$x + 2y - 3z = -7$ $2x - y + z = 13$ $3x + 4y - 2z = 15$

3.2 Problems Involving Simultaneous Equations

Many real life problems can be solved using simultaneous equations. For example:

Four people are training for the triathlon. They use three different circuits:

The swimming circuit is x km long.
The cycling circuit is y km long.
The running circuit is z km long.

(i) Lennon does 3 swimming, 6 cycling and 12 running circuits in a week.
He covers 90 km altogether.
Show that $x + 2y + 4z = 30$

(ii) Sandra does 16 swimming, 12 cycling and 4 running circuits in a week.
She covers 120 km altogether.
Show that $4x + 3y + z = 30$

(iii) Martin swims an extra 2 km each time he does a swimming circuit.
Martin cycles 1 km less each time he does a cycling circuit.
Martin runs the normal running circuit.
Martin does 4 swimming, 10 cycling and 12 running circuits in a week.
He covers 104 km altogether.
Show that $2x + 5y + 6z = 53$

(iv) Hence find the value of x, y and z

(v) Nicola wants to cover 77 km in a week.
She does 7 running circuits and an equal number of swimming and cycling circuits.
How many circuits did she do altogether?

Solution
(i) $3x + 6y + 12z = 90$
You can divide all these terms by 3 to get
$x + 2y + 4z = 30$ (1)

(ii) $16x + 12y + 4z = 120$
You can divide all these terms by 4 to get
$4x + 3y + z = 30$ (2)

(iii) Each swimming circuit $= x + 2$
Each cycling circuit $= y - 1$
Each running circuit $= z$

So $4(x + 2) + 10(y - 1) + 12z = 104$
$4x + 8 + 10y - 10 + 12z = 104$
$4x + 10y + 12z - 2 = 104$
$4x + 10y + 12z = 104 + 2$
$4x + 10y + 12z = 106$

You can divide all these terms by 2 to get
$2x + 5y + 6z = 53$ (3)

(iv) Multiplying (1) by 4 and subtracting (2) gives:

$4x + 8y + 16z = 120$
$-\ 4x + 3y + \quad z = 30$
$\overline{\quad\quad 5y + 15z = 90}$ (4)

Multiplying (1) by 2 and subtracting (3) gives:

$2x + 4y + 8z = 60$
$-\ 2x + 5y + 6z = 53$
$\overline{\quad\quad -y + 2z = 7}$ (5)

Multiplying (5) by 5 and adding (4) gives:

$-5y + 10z = 35$
$+\ \ 5y + 15z = 90$
$\overline{\quad\quad 25z = 125}$
$z = 5$

Substituting $z = 5$ into (4) gives
$5y + 15z = 90$
$5y + 75 = 90$
$5y = 90 - 75 = 15$
$y = 3$

Substituting $z = 5$ and $y = 3$ into (1) gives
$x + 2y + 4z = 30$
$x + 6 + 20 = 30$
$x = 30 - 6 - 20$
$x = 4$

You can check your answers by substituting $x = 4$, $y = 3$ and $z = 5$ into (2):

$4x + 3y + z = 30$
$16 + 9 + 5 = 30$

(v) Swimming circuit = 4 km
Cycling circuit = 3 km
Running circuit = 5 km

7 running circuits = 7 × 5 = 35 km
So there are 77 – 35 = 42 km left to cover
One swimming + one cycling = 4 + 3 = 7 km
So there must be 42 ÷ 7 = 6 swimming and cycling circuits
Total number altogether: 6 + 6 + 7 = 19 circuits

Exercise 3B:

1. A shop sells rulers at x pence each, pens at y pence each and pencils at z pence each.
Cadence buys 12 rulers, 8 pens and 20 pencils for £14
(i) Show that $3x + 2y + 5z = 350$
Lily buys 35 rulers, 15 pens and 30 pencils for £32.25
(ii) Show that $7x + 3y + 6z = 645$
David buys 15 rulers, 12 pens and 24 pencils for £18
(iii) Show that $5x + 4y + 8z = 600$
(iv) Hence find the values of x, y and z
(v) Carly buys 40 of each item and gets 15% discount. How much does she pay altogether?

2. In a computer game you get awarded gold, silver or bronze medals.
A gold medal earns x points.
A silver medal earns y points.
A bronze medal earns z points.
Leslie gets 20 gold, 10 silver and 15 bronze medals. He scores 255 points.

(i) Show that $4x + 2y + 3z = 51$
Audrey gets 12 gold, 16 silver and 8 bronze medals. She scores 200 points.

(ii) Show that $3x + 4y + 2z = 50$
Dylan gets 15 gold, 5 silver and 20 bronze medals. He scores 205 points.

(iii) Show that $3x + y + 4z = 41$

(iv) Hence find the values of x, y and z

(v) Siobhain gets twice as many gold medals as silver and the same number of silver and bronze medals. She gets 672 points altogether. How many medals did she get altogether?

3. There are 3 types of room in a hotel.
Family rooms cost £x per night.
Double rooms cost £y per night.
Single rooms cost £z per night.
On Wednesday 12 family, 6 double and 18 single rooms were occupied. The total cost was £4170.

(i) Show that $2x + y + 3z = 695$
On Thursday 30 family, 10 double and 25 single rooms were occupied. The total cost was £8125.

(ii) Show that $6x + 2y + 5z = 1625$
On Fridays the cost of a family room is increased by £20 and the cost of a double room is increased by £15. The cost of a single room is unchanged.
On Friday 16 family, 12 double and 24 single rooms were occupied. The total cost was £6540.

(iii) Show that $4x + 3y + 6z = 1510$

(iv) Hence find the values of x, y and z

(v) A tourist company needs 4 family, 6 double and 3 single rooms. How much would they save booking a Thursday rather than a Friday.

4. A shop sells shirts at £x each.
It sells trousers at £y each.
It sells jackets at £z each

In a sale Eoin bought a shirt at 20% discount, a pair of trousers at 10% discount and a jacket at 40% discount. These cost £97 altogether.

(i) Show that $8x + 9y + 6z = 970$
In a different sale Padraig bought a shirt at 40% discount, a pair of trousers at 20% discount and a jacket at 20% discount. These cost £108 altogether.

(ii) Show that $3x + 4y + 4z = 540$
In a different sale Trevor bought a shirt at 10% discount, a pair of trousers at 15% discount and a jacket at 25% discount. These cost £111 altogether.

(iii) Show that $18x + 17y + 15z = 2220$

(iv) Hence find the values of x, y and z

(v) When Kevin went to the shop shirts and trousers were on sale at half price.
He bought a shirt, a pair of trousers and a jacket for £83.50
How much percentage discount did he get for the jacket?

5. A shop hires out Blue Ray discs at £x each, DVDs at £y each and Box Sets at £z each.
Eleanor hired 8 Blue Ray discs, 4 DVDs and 12 Box Sets in August for £137

(i) Show that $200x + 100y + 300z = 3425$
Leah hired 15 Blue Ray discs, 20 DVDs and 10 Box Sets from September to December for £247.50

(ii) Show that $30x + 40y + 20z = 495$
In a sale the Blue Ray discs were hired out at 50p off each disc and the Box Sets were hired out at a 10% reduction.
Sarah hired 4 Blue Ray discs, 2 DVDs and 3 Box Sets in the sale for £48.35

(iii) Show that $80x + 40y + 54z = 1007$

(iv) Hence find the values of x, y and z

(v) Claire hired a number of Blue Ray discs, DVDs and Box Sets in the ratio 3:2:4 in the sale for £141.90. How many Box Sets did she hire?

6. Three boys collect model cars, trains and buses.
Each model car weighs x grams.
Each model train weighs y grams.
Each model bus weighs z grams.

Quinn has 6 cars, 8 trains and 20 buses. Their total weight is 28.6 kg.

(i)   Show that $3x + 4y + 10z = 14300$

Arthur has 10 cars, 5 trains and 10 buses. Their total weight is 19.45 kg.

(ii)   Show that $2x + y + 2z = 3890$

Martin has 12 cars, 15 trains and 3 buses. Their total weight is 24.06 kg.

(iii) Show that $4x + 5y + z = 8020$

(iv)   Hence find the values of x, y and z

(v)   Nicola has a number of cars, trains and buses in the ratio 7:8:5. Their total weight is 48.45kg. How many of each vehicle does she have?

7. Teams play hockey matches in a tournament. Each team scores x points for every game won. Each team scores y points for every game drawn. Each team loses z points for every game lost. The table below shows the results and points for four of the teams. Use it to work out the total number of points Larne scored.

Team	Win	Draw	Lose	Total Points
Portstewart	9	6	3	75
Ardglass	4	10	6	46
Coagh	4	12	8	48
Larne	7	9	5	

8. The equation of a curve is $y = ax^3 + bx^2 + cx + 4$
The curve passes through $(-2, -14)$, $(2, -2)$ and $(6, 298)$.
Prove that $(-5, -296)$ lies on this curve.

3.3 One Linear And One Quadratic Equation

When there are two different variables you can eliminate one of them by replacing it in terms of the other variable in one of the equations. For example:

1. Solve $y = x - 3$ and $x^2 - 3y^2 = 13$

You must first replace either x or y in one of the equations. It is probably easier to replace x in the second equation as follows:

$$y = x - 3$$
$$\text{So} \quad x - 3 = y$$
$$x = y + 3$$

The second equation becomes:
$$(y + 3)^2 - 3y^2 = 13$$
This gives:
$$y^2 + 6y + 9 - 3y^2 = 13$$
This simplifies to:
$$-2y^2 + 6y - 4 = 0$$
You can rewrite this with the y^2 term first and positive as follows:
$$2y^2 - 6y + 4 = 0$$
You can divide all the terms by 2 to get:
$$y^2 - 3y + 2 = 0$$
You can then factorise this to get:
$$(y - 1)(y - 2) = 0$$
Either $\quad y - 1 = 0$
$$y = 1$$
or $\quad y - 2 = 0$
$$y = 2$$
You must now substitute these values into one of the equations to find the values of x
It is easier to use $\quad x = y + 3$
When $y = 1$ so $\quad x = 1 + 3 = 4$
When $y = 2$ so $\quad x = 2 + 3 = 5$
The answer is $x = 4$, $y = 1$ and $x = 5$ $y = 2$

....

2. Find the coordinates of the points where the straight line $2y + 3x = 1$ cuts the curve $x^2 + xy = -3$

It is probably easier to replace y in the second equation as follows:
$$2y + 3x = 1$$
$$\text{So} \quad 2y = 1 - 3x$$
$$y = \frac{1}{2} - \frac{3}{2}x$$

The second equation becomes:
$$x^2 + x\left(\frac{1}{2} - \frac{3}{2}x\right) = -3$$

This gives:
$$x^2 + \frac{1}{2}x - \frac{3}{2}x^2 = -3$$

This simplifies to:
$$-\frac{1}{2}x^2 + \frac{1}{2}x + 3 = 0$$

You can multiply all the terms by 2 to get:
$$-x^2 + x + 6 = 0$$
You can rewrite this with the x^2 term first and positive as follows:
$$x^2 - x - 6 = 0$$
You can then factorise this to get:
$$(x + 2)(x - 3) = 0$$

Either $\quad x + 2 = 0$

$\qquad\qquad x = -2$

or $\qquad x - 3 = 0$

$\qquad\qquad x = 3$

You must now substitute these values into one of the equations to find the values of y

It is easier to use $2y + 3x = 1$

When $x = -2$ so

$\qquad 2y - 6 = 1$

$\qquad\quad 2y = 1 + 6$

$\qquad\quad 2y = 7$

$\qquad\quad\ y = 3.5$

When $x = 3$ so

$\qquad 2y + 9 = 1$

$\qquad\quad 2y = 1 - 9$

$\qquad\quad 2y = -8$

$\qquad\quad\ y = -4$

The answer is $(-2, 3.5)$ and $(3, -4)$

1. $2x - y = 1$ and $y^2 - x^2 = 33$
2. $4y - x = 3$ and $xy - y^2 = 6$
3. $3x + 2y = 6$ and $2x^2 + 3xy + 4 = 0$
4. $y - 2x = 10$ and $x^2 - xy = 16$
5. $3y - x = 3$ and $x^2 - y^2 = 27$
6. $2y - x = 3$ and $xy - y^2 = 4$
7. Find the coordinates of the points where the straight line $x + 3y = 1$ cuts the curve $x^2 - 5y^2 = 29$
8. Find the coordinates of the points where the straight line $y - x = 8$ cuts the curve $x^2 + xy = 10$
9. Find the coordinates of the points where the straight line $x - 3y = 5$ cuts the curve $x^2 + y^2 = 5$
10. Find the coordinates of the points where the straight line $2y + 3x = 2$ cuts the curve $y^2 + xy = 8$

CHAPTER 4: TRIGONOMETRY

4.1 Trigonometric Ratios

When turning through an angle, we always start with the positive horizontal axis. Therefore the angle at this line is always taken as 0°. As we turn from the positive horizontal axis in an anti–clockwise direction the angles change from 0° to 360°.

Consider these angles from 0° to 360°:
- All the sines, cosines and tangents of all the angles between 0° and 90° are positive

  Eg sin 30° = 0.5

  $\quad$ cos 45° = 0.707

  $\quad$ tan 60° = 1.732
- However between 90° and 180° only the sines are positive

  Eg sin 100° = 0.985

  $\quad$ cos 145° = −0.819

  $\quad$ tan 160° = −0.364
- Between 180° and 270° only the tangents are positive

  Eg sin 200° = −0.342

  $\quad$ cos 245° = −0.423

  $\quad$ tan 260° = 5.671

- Between 270° and 360° only the cosines are positive

  eg sin 300° = −0.866

  $\quad$ cos 345° = 0.966

  $\quad$ tan 350° = −0.176

We can summarise these in the sketch below:

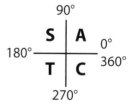

As we turn from the positive horizontal axis in a clockwise direction the angles change from 0° to −360°

Now, consider the angles from −180° to 180°:

We can summarise the sines, cosines and tangents of all these angles in the sketch below:

4.2 Solving Trig Equations Between 0° And 360°

For example:

1. Solve sin $x = 0.7$ for $0° \leq x \leq 360°$
There are two quadrants in which the sin is positive as shown below:

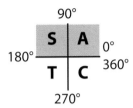

You can find one of the solutions using the inverse sine button on your calculator, ie sin⁻¹ 0.7 = 44.4°
The other angle is in the quadrant between 90° and 180°. You can find it by subtracting 44.4 from 180 as shown below.

44.4°   180 – 44.4 = 135.6°

The answer is 44.4° or 135.6°

....

2. Solve cos $x = -0.345$ for $0° \leq x \leq 360°$
There are two quadrants in which the cos is negative as shown below:

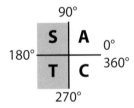

You should solve cos $x = 0.345$ first using the inverse cos button on your calculator, and then use your answer to find the two solutions, ie cos⁻¹ 0.345 = 69.8°
You can find the angle in the quadrant between 90° and 180° by subtracting 69.8 from 180 as shown below.

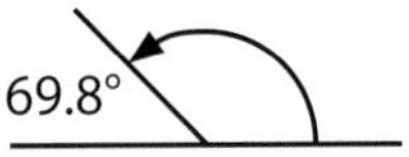

69.8°   180 – 69.8 = 110.2°

You can find the angle in the quadrant between 180° and 270° by adding 69.8 to 180 as shown below

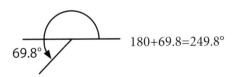

69.8°   180+69.8=249.8°

The answer is 110.2° or 249.8°

3. Solve tan $x = -4$ for $0° \leq x \leq 360°$
There are two quadrants in which the tan is negative as shown below:

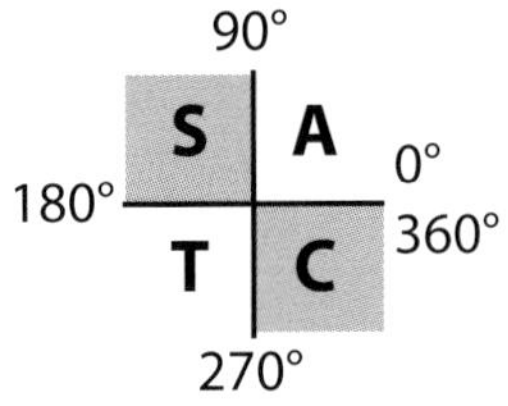

You should solve tan $x = 4$ first using the inverse tan button on your calculator and then use your answer to find the two solutions, ie tan⁻¹ 4 = 75.96°
You can find the angle in the quadrant between 90° and 180° by subtracting 75.96 from 180 as shown below:

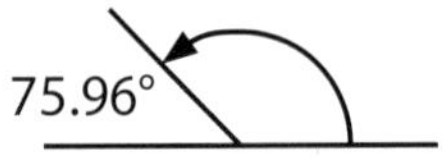

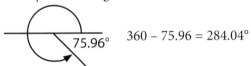

75.96°   180 – 75.96 = 104.04°

You can find the angle in the quadrant between 270° and 360° by subtracting 75.96 from 360 as shown below:

75.96°   360 – 75.96 = 284.04°

The answer is 104.04° or 284.04°

Exercise 4A: *Solve the following where 0° ≤ x ≤ 360°*

1. cos $x = 0.462$

2. sin $x = -0.762$

3. tan $x = 3$

4. sin $x = 0.2$

5. cos $x = -0.75$

6. tan $x = -0.6$

4.3 Solving Trig Equations Between −180° And 180°

For example:

1. Solve sin $x = -0.43$ for $-180° \leq x \leq 180°$
There are two quadrants in which the sin is negative as shown below:

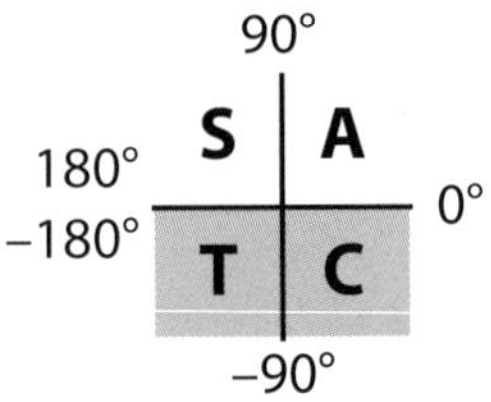

You should solve sin $x = 0.43$ first using the inverse sin button on your calculator and then use your answer to

find the two solutions, ie $\sin^{-1} 0.43 = 25.5°$
The angle in the quadrant between 0° and –90° is then

–25.5°:

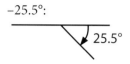

You can find the angle in the quadrant between –90° and –180° by subtracting 25.5 from 180 and changing the sign to a negative as shown below:

180 – 25.5 = 154.5°

The answer is –25.5° or –154.5°

. . . .

2. $\cos x = 0.845$ for $-180° \leq x \leq 180°$
There are two quadrants in which the cos is positive as shown below:

```
              90°
         S  |  A
180°   ─────┼─────  0°
–180°    T  |  C
             –90°
```

You can find one of the solutions using the inverse cos button on your calculator, ie $\cos^{-1} 0.845 = 32.3°$
The other angle is in the quadrant between 0° and –90° and so it will just be –32.3°:

32.3°

The answer is 32.3° or –32.3°

. . . .

3. $\tan x = 0.2$ for $-180° \leq x \leq 180°$
There are two quadrants in which the tan is positive as shown below:

```
              90°
         S  |  A
180°   ─────┼─────  0°
–180°    T  |  C
             –90°
```

You can find one of the solutions using the inverse tan button on your calculator, ie $\tan^{-1} 0.2 = 11.3°$
The other angle is in the quadrant between –90° and –180°
You can find this angle by subtracting 11.3 from 180 and

changing the sign to a negative as shown below:

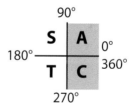

180 – 11.3 = 168.7°

The answer is 11.3° or –168.7°

Exercise 4B: *Solve the following where* $-180° \leq x \leq 180°$

1. $\cos x = 0.6$

2. $\sin x = -0.34$

3. $\tan x = -2$

4. $\sin x = 0.739$

5. $\cos x = -0.349$

6. $\tan x = -1.5$

4.4 More Complex Trig Equations
For example:

1. Solve $\cos 5x = 0.482$ where $0° \leq x \leq 72°$
It is probably easier to let $y = 5x$ and solve $\cos y = 0.482$ as before.
Then you can set up an algebraic equation for x as follows:
First you solve $\cos y = 0.482$
There are two quadrants in which the cos is positive from 0° to 360° as shown below:

```
              90°
         S  |  A
180°   ─────┼─────  0°
         T  |  C    360°
             270°
```

You can find one of the solutions using the inverse cos button on your calculator, ie $\cos^{-1} 0.482 = 61.2°$
The other angle is in the quadrant between 270° and 360°. You can find it by subtracting 61.2 from 360 as shown below:

360 – 61.2 = 298.8°

So $y = 61.2°$ or 298.8°
You can now set up equations for x by substituting $5x = y$
$$5x = 61.2°$$
$$x = 12.24°$$
or
$$5x = 298.8°$$
$$x = 59.76°$$
These angles are both between 0° and 72°
(since 360 ÷ 5 = 72)

2. Solve tan(3x – 10) = –3 where –60° ≤ x ≤ 60°

It is probably easier to let y = 3x – 10 and solve tan y = –3 as before.

Then you can set up an algebraic equation for y as follows:

First you solve tan y = –3

There are two quadrants in which the tan is negative from –180° to 180° as shown below:

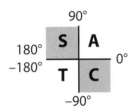

You can find one of the solutions using the inverse tan button on your calculator, ie tan⁻¹ 3 = 71.6°

One angle is in the quadrant between 0° and –90°

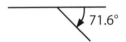

This angle is –71.6°

The other angle is in the quadrant between 90° and 180°

You can find it by subtracting 71.6 from 180 as shown below

71.6°   180 – 71.6 = 108.4°

So y = –71.6° or 108.4°

You can now set up equations for x by substituting 3x – 10 = y

$$3x - 10 = -71.6°$$
$$3x = -71.6° + 10$$
$$3x = -61.6°$$
$$x = -20.5°$$

or

$$3x - 10 = 108.4°$$
$$3x = 108.4° + 10$$
$$3x = 118.4°$$
$$x = 39.5°$$

These angles are both between –60° and 60° (since ±180 ÷ 3 = ±60)

Exercise 4C:

1. Solve sin 3x = 0.467 where 0° ≤ x ≤ 120°

2. Solve cos 2x = –0.16 where 0° ≤ x ≤ 180°

3. Solve tan 4x = 1.3 where 0° ≤ x ≤ 90°

4. Solve sin(x + 40) = –0.96 where 0° ≤ x ≤ 360°

5. Solve cos(x – 25) = 0.12 where 0° ≤ x ≤ 360°

6. Solve tan(x – 10) = –0.5 where –180° ≤ x ≤ 180°

7. Solve cos(⅓x – 47) = 0.12 where –540° ≤ x ≤ 540°

8. Solve sin 2x = 0.5 where 0° ≤ x ≤ 180°

9. Solve cos(3x – 10) = –0.2 where –120° ≤ x ≤ 120°

10. Solve tan(x – 25) = 2 where –180° ≤ x ≤ 180°

11. Solve tan(5x + 22) = –0.34 where –36° ≤ x ≤ 36°

12. Solve sin(½x + 15) = –0.174 where 0° ≤ x ≤ 720°

CHAPTER 5: SOLUTION OF TRIANGLES

You can find missing sides or angles in triangles by using:
- **Pythagoras' theorem** and **trigonometry** for right–angled triangles
- the **sine rule** or the **cosine rule** for triangles that are not right–angled

5.1 Pythagoras' Theorem

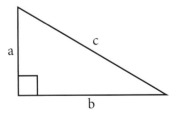

With reference to the diagram, Pythagoras' theorem states that $a^2 + b^2 = c^2$

For example:

1. Triangle PQR is right–angled at Q.

$$PQ = 7cm$$
$$QR = 4cm$$

Find the length of PR.

P

7 cm

Q   4 cm   R

Let PR = x
$c^2 = a^2 + b^2$
$x^2 = 7^2 + 4^2$
$x^2 = 65$
$x = \sqrt{65} = 8.06$ cm

....

2. Triangle PQR is right–angled at Q.

$$PQ = 15.8\text{cm}$$
$$PR = 24.6\text{cm}$$

Find the length of QR.

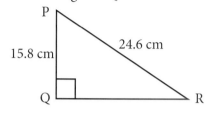

Let QR = x

$$x^2 + 15.8^2 = 24.6^2$$
$$x^2 = 24.6^2 - 15.8^2$$
$$x^2 = 355.52$$
$$x = \sqrt{355.52} = 18.9 \text{ cm}$$

5.2 Trigonometry

The three trig ratios in a right–angled triangle are defined as:

$$\sin x = \frac{\text{Opposite}}{\text{Hypotenuse}}$$

$$\cos x = \frac{\text{Adjacent}}{\text{Hypotenuse}}$$

$$\tan x = \frac{\text{Opposite}}{\text{Adjacent}}$$

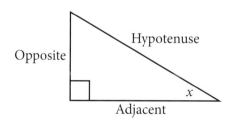

For example:
1. Triangle PQR is right–angled at Q.
$$PQ = 7 \text{ cm}$$
$$QR = 4 \text{ cm}$$
Find the size of ∠PRQ

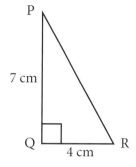

$$\tan x = 7/4$$
$$\tan x = 1.75$$
$$x = \tan^{-1} 1.75 = 60.3°$$

....

2. Triangle ABC is right–angled at B.
$$AB = 15 \text{ cm}$$
$$∠BAC = 53°$$
Find the length of AC

Let AC = x

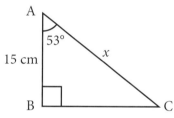

$$\cos 53 = \frac{15}{x}$$
$$x \cos 53 = 15$$
$$x = \frac{15}{\cos 53}$$
$$x = 24.9 \text{ cm}$$

Exercise 5A:

1. Triangle ABC is right–angled at B.
$$AB = 7.8\text{cm}$$
$$BC = 6\text{cm}$$
Find the length of AC

2. Triangle ABC is right–angled at B.
$$BC = 4.6\text{cm}$$
$$AC = 10\text{cm}$$
Find the size of ∠BAC

3. Triangle ABC is right–angled at B.
$$AC = 3.2\text{cm}$$
$$BC = 2.7\text{cm}$$
Find the length of AB

Exercise 5A...

4. Triangle ABC is right–angled at B.

AC = 15cm

∠ACB = 40°

Find the length of BC

5. Triangle ABC is right–angled at B.

AB = 4cm

BC = 5cm

Find the size of ∠BAC

6. Triangle ABC is right–angled at B.

AB = 9.2cm

BC = 6.8cm

Find the length of AC

7. Triangle ABC is right–angled at B.

AB = BC

AC = 16cm

Find the perimeter of ABC

8. Triangle ABC is right–angled at B.

BC = 3cm

∠BAC = 25°

Find the length of AB

5.3 The Sine Rule

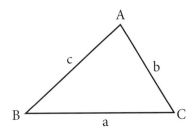

With reference to the diagram, the sine rule states:

$$\frac{\sin A}{a} = \frac{\sin B}{b} = \frac{\sin C}{c}$$

You use the sine rule when:
• there is no right angle;
• there are two sides and two angles, including the unknown side/angle.

For example:

1. ABC is a triangle in which AB = 23 cm, AC = 18 cm and ∠ACB = 53°

Find the size of ∠ABC

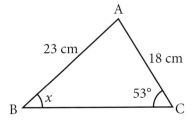

Using the sine rule:

$$\frac{\sin x}{18} = \frac{\sin 53}{23}$$

You can cross multiply to get:

$$23 \sin x = 18 \sin 53$$

So

$$\sin x = \frac{18 \sin 53}{23}$$

$$\sin x = 0.625$$

$$x = \sin^{-1} 0.625 = 38.7°$$

. . . .

2. ABC is a triangle in which BC = 14.5cm, ∠ACB = 73° and ∠BAC = 62°   Find the size of AC

Let AC = x

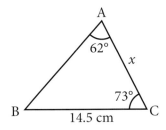

You must first work out ∠ABC

$$73° + 62° = 135°$$

So

$$∠ABC = 180° – 135°$$

$$= 45°$$

You can then use the sine rule to get:

$$\frac{\sin 45}{x} = \frac{\sin 62}{14.5}$$

You can cross multiply to get:

$$x \sin 62 = 14.5 \sin 45$$

So

$$x = \frac{14.5 \sin 45}{\sin 62}$$

$$x = 11.6 \text{ cm}$$

5.4 Cosine Rule

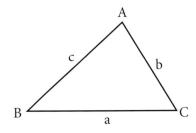

With reference to the diagram, the cosine rule states:
$a^2 = b^2 + c^2 - 2bc \cos A$

You use the cosine rule when:
- there is no right angle;
- there are 3 sides and 1 angle, including the unknown side/angle

For example:

1. ABC is a triangle in which AB = 4.7 cm, BC = 5.3 cm and $\angle ABC = 34°$

Find AC

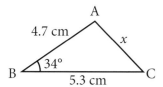

Using the cosine rule

$$x^2 = 4.7^2 + 5.3^2 - 2 \times 4.7 \times 5.3 \cos 34$$
$$x^2 = 8.877$$
$$x = \sqrt{8.877}$$

So $\qquad$ AC = 2.98 cm

---

2. ABC is a triangle in which AB = 14 cm, AC = 13 cm and BC = 19 cm

Find the size of $\angle BAC$

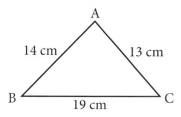

Using the cosine rule

$$19^2 = 14^2 + 13^2 - 2 \times 14 \times 13 \cos A$$
$$361 = 365 - 364 \cos A$$
$$364 \cos A = 365 - 361$$
$$364 \cos A = 4$$
$$\cos A = \frac{4}{364}$$
$$\cos A = 0.011$$

So $\qquad$ $A = \cos^{-1} 0.011 = 89.4°$

Exercise 5B:

1. ABC is a triangle in which AB = 9.8 cm, AC = 8.4 cm and $\angle ACB = 64°$

Find the size of $\angle ABC$

Exercise 5B...

2. ABC is a triangle in which AB = 7.2 cm, BC = 7.4 cm and AC = 6.8 cm

Find the size of $\angle BAC$

3. ABC is a triangle in which BC = 7.5 cm, $\angle ACB = 64°$ and $\angle BAC = 52°$

Find AB

4. ABC is a triangle in which AB = 6.2 cm, BC = 5.9 cm and $\angle ABC = 47°$

Find AC

5. ABC is a triangle in which AB = 45 cm, AC = 38 cm and BC = 42 cm

Find the size of $\angle ABC$

6. ABC is a triangle in which BC = 6.4 cm, $\angle ACB = 81°$ and $\angle BAC = 53°$

Find AC

7. ABC is a triangle in which AB = 4.2 cm, AC = 2.8 cm and $\angle BAC = 126°$

Find BC

8. ABC is a triangle in which AB = 5.4 cm, BC = 7.2 cm and $\angle BAC = 50°$

Find the size of $\angle ABC$

5.5 Area of a Triangle

You can find the area of a triangle using either:
- Area = ½ base × perpendicular height
  (should normally be used in a right–angled triangle); or
- Area = ½ bc sin A
  (should be used when it is not a right–angled triangle)

For example:

1. Triangle ABC is right–angled at B.
$$AC = 12\text{cm}$$
$$BC = 7\text{cm}$$

Find the area of ABC

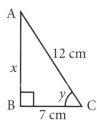

You can use either area formula to answer this question.

- **Method 1:**

You need to find the length of AB using Pythagoras' Theorem:

$$x^2 + 7^2 = 12^2$$
$$x^2 + 49 = 144$$

$$x^2 = 95$$
$$x = \sqrt{95}$$
$$x = 9.75 \text{ cm}$$

Then you can use Area = ½ base × perpendicular height

$$\text{Area} = \frac{1}{2} \times 7 \times 9.75$$
$$\text{Area} = 34.1 \text{ cm}^2$$

- **Method 2:**

You need to find ∠ACB using trigonometry:

$$\cos y = \frac{7}{12}$$
$$\cos y = 0.583$$
$$y = \cos^{-1} 0.583 = 54.3°$$

Then you can use Area = ½ bc sin A

$$\text{Area} = \frac{1}{2} \times 7 \times 12 \sin 54.3°$$
$$\text{Area} = 34.1 \text{ cm}^2$$

2. ABC is a triangle in which AB = 7cm, AC = 5.8 cm and ∠ACB = 38°.
Find the area of ABC

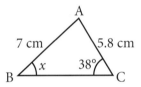

In this case, you have to use Area = ½ bc sin A
You first need to find ∠ABC first using the sine rule:

$$\frac{\sin x}{5.8} = \frac{\sin 38}{7}$$

You can cross multiply to get:

$$7 \sin x = 5.8 \sin 38$$

So
$$\sin x = \frac{5.8 \sin 38}{7}$$

$$\sin x = 0.510$$
$$x = \sin^{-1} 0.510 = 30.7°$$

You can then find ∠BAC:

$$38° + 30.7° = 68.7°$$

So
$$∠BAC = 180° - 68.7° = 111.3°$$

You can now find the area using Area = ½ bc sin A

$$\text{Area} = \frac{1}{2} \times 7 \times 5.8 \sin 111.3°$$
$$\text{Area} = 18.9 \text{ cm}^2$$

Exercise 5C:

1. Triangle ABC is right–angled at B.
$$BC = 6\text{cm}$$
$$∠BAC = 34°$$
Find the area of ABC

2. ABC is a triangle in which AB = 6.4 cm,
∠ABC = 50° and ∠ACB = 36°
Find the area of ABC

3. Triangle ABC is right–angled at B.
$$AB = 2.7\text{cm}$$
$$AC = 4.6\text{cm}$$
Find the area of ABC

4. ABC is a triangle in which AB = 8.2 cm,
AC = 9.4 cm and ∠ABC = 37°
Find the area of ABC

5. ABC is an isosceles triangle where AC = BC.
$$∠ABC = 52°$$
$$AB = 9.4\text{cm}$$
Find the area of ABC

6. Triangle ABC is right–angled at B.
$$AC = 5.3\text{cm}$$
$$∠ACB = 54°$$
Find the area of ABC

7. ABC is a triangle in which AB = 8.2 cm,
AC = 7.4 cm and ∠ACB = 134°
Find the area of ABC

8. Triangle ABC is right–angled at B.
$$AC = 7.2\text{cm}$$
$$AB = 4\text{cm}$$
Find the area of ABC

Exercise 5D:

1. ABCD is a quadrilateral in which AD = 11.5 cm, CD = 9.8 cm and BC = 8 cm
Angle ABC = 46° and angle BAC = 78°

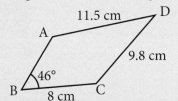

Find (i) AC
  (ii) angle ADC
  (iii) the area of ABCD

2. ABCD is a quadrilateral in which AD = 7.5 cm, BD = 6 cm and DC = 5 cm
Angle ABD= 84° and angle BDC = 56°

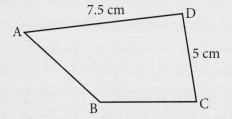

Find (i) angle BAD
  (ii) BC
  (iii) the shortest distance from D to BC

3. ABCD is a quadrilateral in which AD = 7.9 cm and CD = 10.3 cm
Angle BAD = 90°, ADB = 35° and angle BCD = 48°

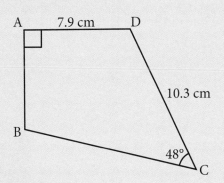

Find (i) BD
  (ii) angle ADC
  (iii) the area of ABCD

4. B is on a bearing of 036° from A and 8.4 km from A.
C is on a bearing of 147° from A and 13.2 km from A.
Find (i) BC
  (ii) the bearing of C from B

5. ABCD is a quadrilateral in which AD = 7.4 cm and CD = 8.6 cm
Angle ADB = 50°
The area of ABD is 23.4 cm²
The area of ABCD is 56 cm²

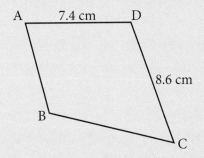

Find (i) BD
  (ii) angle ADC
  (iii) the perimeter of ABCD

CHAPTER 6: DIFFERENTIATION

6.1 Rules of Differentiation

You can differentiate any algebraic function using the following rules:

- Powers of x other than 1:
  1. Multiply by the power
  2. Take 1 off the power
  $y = kx^n$ (where k is any constant)

$$\frac{dy}{dx} = nkx^{n-1}$$

- $y = kx$ (where k is any constant)

$$\frac{dy}{dx} = k$$

- $y = k$ (where k is any constant)

$$\frac{dy}{dx} = 0$$

For example, differentiate the following with respect to x:

1. $y = 4x^3 + 6x^2 - 7x - 5$

Then $\frac{dy}{dx} = 12x^2 + 12x - 7$

. . . .

2. $y = 2x^3 + mx^2 - nx + 3$

You treat m and n as constants

Then $\frac{dy}{dx} = 6x^2 + 2mx - n$

Exercise 6A: *Differentiate the following with respect to x:*

1. $y = 6x + 2$
2. $y = 2x^2 - 3x + 4$
3. $y = 3x^2 + x - 5$
4. $y = 2x^3 + 3x^2 - 5x + 2$
5. $y = 4x^3 - 6x^2 + x - 7$
6. $y = 2 + 3x - x^2 + 7x^3$
7. $y = 8 - 3x + 2x^2 - x^3$
8. $y = x^4 - 2x^2 + 7$
9. $y = 6x^3 - 3x + 5$
10. $y = 4x^2 - x^3 + 2x^4$

Exercise 6A...

11. $y = 2x^5 - 3x^4 + 2x^3 - 6x^2 + x$
12. $y = 4x^3 - 7x^2 + 2$
13. $y = 6x^4 - 2x^2 + x$
14. $y = 4 + 3x - x^3$
15. $y = 2x^3 - 4x^2 - 7x$
16. $y = 6 - 2x + 3x^2 - x^3$
17. $y = 2x^3 - x^2 + 4x - 3$
18. $y = 4x^2 - 5x + 3$
19. $y = ax^2 + bx$
20. $y = 8x^3 - vx$
21. $y = 9x^2 + tx - w$

6.2 Differentiation With Fractional Coefficients

You follow the same rules. But remember that when you multiply a fraction by a whole number you only multiply the **numerator** of the fraction by the whole number

eg $\frac{2}{5} \times 6 = \frac{12}{5}$

For example, differentiate the following with respect to x:

1. $y = \frac{1}{8}x^4 - \frac{3}{5}x^2 + 7x$

Then $\frac{dy}{dx} = \frac{4}{8}x^3 - \frac{6}{5}x + 7$

You should cancel the fraction if possible.

Final answer is $\frac{dy}{dx} = \frac{1}{2}x^3 - \frac{6}{5}x + 7$

. . . .

2. $y = 8 + x - \dfrac{2}{9}x^3 + \dfrac{3}{4}x^8$

Then $\dfrac{dy}{dx} = 1 - \dfrac{6}{9}x^2 + \dfrac{24}{4}x^7$

You should then cancel the fraction.

Final answer is $\dfrac{dy}{dx} = 1 - \dfrac{2}{3}x^2 + 6x^7$

Exercise 6B: *Differentiate the following with respect to x:*

1. $y = \dfrac{1}{4}x^3 - \dfrac{2}{5}x^2 + x$

2. $y = 2x^3 - \dfrac{1}{5}x^2 + \dfrac{3}{4}x$

3. $y = 4x^3 - \dfrac{2}{7}x^2 + \dfrac{1}{3}x$

4. $y = \dfrac{1}{5}x^3 - \dfrac{3}{2}x^2 + 4x$

5. $y = \dfrac{2}{9}x^3 + \dfrac{1}{4}x^2 - 3$

6. $y = \dfrac{3}{4}x^2 + \dfrac{1}{5}x - 6$

7. $y = 3x^3 - \dfrac{3}{4}x^2 + x$

8. $y = 4 - \dfrac{1}{5}x + \dfrac{5}{2}x^2$

9. $y = \dfrac{7}{3}x^3 - \dfrac{2}{5}x^2 + \dfrac{1}{4}x$

10. $y = 3 + \dfrac{1}{2}x - \dfrac{5}{6}x^2 + \dfrac{7}{3}x^3$

6.3 Differentiation With The x Term On The Denominator

You must rewrite the x term using the rule $\dfrac{1}{x^n} = x^{-n}$

For example, differentiate the following with respect to x:

1. $y = \dfrac{8}{x^3}$

Rewriting gives $y = 8x^{-3}$

Then $\dfrac{dy}{dx} = -24x^{-4}$

You should then rewrite the answer with a positive index.

Final answer $\dfrac{dy}{dx} = \dfrac{-24}{x^4}$

···

2. $y = \dfrac{9}{8x^2}$

Rewriting gives $y = \dfrac{9}{8}x^{-2}$

Then $\dfrac{dy}{dx} = \dfrac{-18}{8}x^{-3}$

You should then cancel.

So $\dfrac{dy}{dx} = \dfrac{-9}{4}x^{-3}$

Final answer $\dfrac{dy}{dx} = \dfrac{-9}{4x^3}$

···

3. $y = \dfrac{1}{3x^6}$

Rewriting gives $y = \dfrac{1}{3}x^{-6}$

Then $\dfrac{dy}{dx} = \dfrac{-6}{3}x^{-7}$

You should then cancel.

So $\dfrac{dy}{dx} = -2x^{-7}$

Final answer $\dfrac{dy}{dx} = \dfrac{-2}{x^7}$

Exercise 6C: *Differentiate the following with respect to x:*

1. $y = \dfrac{6}{x^2}$ **6.** $y = \dfrac{2}{7x^3}$

2. $y = \dfrac{4}{x^3}$ **7.** $y = \dfrac{4}{x^3}$

3. $y = \dfrac{5}{x^4}$ **8.** $y = \dfrac{3}{x^5}$

4. $y = \dfrac{3}{2x^7}$ **9.** $y = \dfrac{1}{2x^3}$

5. $y = \dfrac{5}{4x^2}$ **10.** $y = \dfrac{2}{4x^2}$

6.4 Further Differentiation

For example, differentiate the following with respect to x:

$$y = \frac{2}{9}x^3 + \frac{1}{4}x^2 - \frac{8}{x^3}$$

Rewriting gives $y = \frac{2}{9}x^3 + \frac{1}{4}x^2 - 8x^{-3}$

Then $\quad \frac{dy}{dx} = \frac{6}{9}x^2 + \frac{2}{4}x + 24x^{-4}$

You should then cancel to get:

$$\frac{dy}{dx} = \frac{2}{3}x^2 + \frac{1}{2}x + 24x^{-4}$$

You should then rewrite the answer with a positive index.

Final answer $\quad \frac{dy}{dx} = \frac{2}{3}x^2 + \frac{1}{2}x + \frac{24}{x^4}$

Exercise 6D: Differentiate the following with respect to x:

1. $y = 6x^3 + \frac{2}{x^4}$

2. $y = 2 + \frac{3}{x^5}$

3. $y = 2x^5 + 3x - \frac{4}{x^2}$

4. $y = \frac{2}{3}x^6 - 4 + \frac{2}{3}x^2$

5. $y = \frac{3}{4}x^4 + x - \frac{3}{4x^3}$

6. $y = \frac{7}{2}x^2 - \frac{2}{7x^2}$

7. $y = 4x^3 + x^2 - \frac{1}{x^4}$

8. $y = 2x^2 + 6 + \frac{1}{3x^2}$

9. $y = 4x^2 - 7x + \frac{2}{x^2}$

10. $y = 5x + \frac{2}{3x^4}$

11. $y = 7x^2 - 8 + \frac{2}{x^6}$

12. $y = 3x^4 + \frac{2}{x^5} - 3x$

13. $y = \frac{4}{3}x^6 - \frac{3}{4x^6}$

14. $y = 2x^4 + 5x - \frac{2}{3x^5}$

15. $y = 3x^5 + \frac{2}{3}x^3 - \frac{3}{2x^2}$

6.5 Differentiation of $\frac{dy}{dx}$

You can differentiate $\frac{dy}{dx}$ to get $\frac{d^2y}{dx^2}$

You use the same rules as before. For example:

1. $y = 2x^3 + 3x^2 - 5x + 2$

Find $\frac{dy}{dx}$ and $\frac{d^2y}{dx^2}$

$$\frac{dy}{dx} = 6x^2 + 6x - 5$$

$$\frac{d^2y}{dx^2} = 12x + 6$$

2. $y = 3x^4 + \frac{2}{x^5} - 3x$

Find $\frac{dy}{dx}$ and $\frac{d^2y}{dx^2}$

$$y = 3x^4 + 2x^{-5} - 3x$$

$$\frac{dy}{dx} = 12x^3 - 10x^{-6} - 3$$

You should leave the index as a negative power to make it easier to differentiate again.

$$\frac{d^2y}{dx^2} = 36x^2 + 60x^{-7}$$

Exercise 6E: Find $\frac{dy}{dx}$ and $\frac{d^2y}{dx^2}$ in the following:

1. $y = 3x^4 - 7x^3 + 2x^2$

2. $y = 6x + 2x^2 - 7$

3. $y = 2x^3 + 7x^2 - 5x + 4$

4. $y = 3x + \frac{2}{x^2}$

5. $y = 2 - x - x^2$

6. $y = 4x - 2x^2 + 1$

7. $y = 6x^2 + \frac{1}{x^2}$

8. $y = \frac{2}{3}x^3 - 4x - 2$

9. $y = \frac{3}{5}x^{10} + \frac{1}{5x^2}$

10. $y = \frac{3}{2}x^2 + \frac{2}{3x^2}$

11. $y = 4x^3 - 4 - \frac{2}{x^3}$

12. $y = 6x^2 - \frac{1}{6x^3}$

13. $y = 3x^2 - x - \frac{1}{x^2}$

14. $y = 2x + 3x^2 - \frac{4}{x^3}$

15. $y = 6x + \frac{2}{5x^4}$

CHAPTER 7: TANGENTS AND NORMALS

7.1 Gradient

You can find the gradient of the tangent to any curve $y = f(x)$ at any point by:

Differentiating y to get $\frac{dy}{dx}$.

Substituting the x coordinate of the point into $\frac{dy}{dx}$.

For example, find the gradient of the tangent to the following curves at the following points:

1. $y = 4x^3 + 2x^2 - 6x + 4$ at the point where $x = 2$

$\frac{dy}{dx} = 12x^2 + 4x - 6$

Substituting $x = 2$ gives:
$48 + 8 - 6 = 50$
Answer: Gradient = 50

2. $y = \frac{2}{3}x^2 + 7x + \frac{3}{x^2}$ at $(-1, -3\frac{1}{3})$

$\frac{dy}{dx} = \frac{4}{3}x + 7 - \frac{6}{x^3}$

Substituting $x = -1$ gives:

$\frac{-4}{3} + 7 + 6 = 11\frac{2}{3}$

Answer: Gradient = $11\frac{2}{3}$

Exercise 7A: *Find the gradient of the tangent to the following curves at the following points:*

1. $y = 2x^2 - 7x + 3$ at the point where $x = 2$

2. $y = 3x + 4x^2 - 1$ at the point where $x = 3$

3. $y = 6 - x + x^2$ at the point where $x = -2$

4. $y = 2x^3 + 4x - 5$ at the point where $x = -1$

5. $y = \frac{2}{5}x^2 + \frac{1}{3}x + 2$ at the point where $x = 4$

6. $y = \frac{3}{4}x^2 - x - 4$ at the point where $x = 2$

7. $y = \frac{2}{3}x^3 + \frac{1}{2}x^2 + x$ at the point where $x = -3$

Exercise 7A...

8. $y = \frac{4}{x^2}$ at $(2, 1)$

9. $y = \frac{1}{x^3}$ at $(2, \frac{1}{8})$

10. $y = 6x + \frac{1}{x^2}$ at $(-2, -11\frac{3}{4})$

11. $y = \frac{10}{x^4}$ at the point where $x = 2$

12. $y = (x + 2)(3x - 1)$ at $(2, 20)$

13. $y = (2x - 1)^2$ at $(-2, 25)$

14. $y = 3x^2 + x - \frac{1}{x^2}$ at the point where $x = 4$

15. $y = 4 - \frac{2}{x^3}$ at the point where $x = -2$

You can also find x and y when the gradient is known. The rules are as follows:

- Differentiate y to get $\frac{dy}{dx}$
- Put $\frac{dy}{dx}$ equal to the gradient
- Solve your equation to find x
- Substitute x into y to find y

For example:
Find where the gradient of the tangent to the curve $y = 2x^2 - 7x + 8$ is 5

$$\frac{dy}{dx} = 4x - 7$$
$$4x - 7 = 5$$
$$4x = 12$$
$$x = 3$$

$$y = 2x^2 - 7x + 8$$
$$y = 2(3)^2 - 7(3) + 8$$
$$y = 18 - 21 + 8$$
$$y = 5$$

Answer (3,5)

Exercise 7B:

1. Find where the gradient of the tangent to the curve $y = 2x^2 - 5x + 3$ is 7

2. Find where the gradient of the tangent to the curve $y = 8 + x - 2x^2$ is 9

3. Find where the gradient of the tangent to the curve $y = 4x^2 - 5x + 3$ is 27

4. Find where the gradient of the tangent to the curve $y = 10 - 2x + x^2$ is –4

5. Find where the gradient of the tangent to the curve $y = \frac{3}{2}x^2 - 4x + 3$ is 2

6. Find where the gradient of the tangent to the curve $y = 6 - 2x - 3x^2$ is 10

7. Find where the gradient of the tangent to the curve $y = (x + 4)(2x - 1)$ is –5

8. Find where the gradient of the tangent to the curve $y = (3x - 5)^2$ is 24

9. Find where the gradient of the tangent to the curve $y = x^2 - 10x + 8$ is –18

10. Find where the gradient of the tangent to the curve $y = 4 + 5x - 2x^2$ is –3

11. Find where the gradient of the tangent to the curve $y = 4x^2 - 5x + 7$ is parallel to the x axis

12. Find where the gradient of the tangent to the curve $y = \frac{5}{2}x^2 + 2x - 1$ is horizontal

13. Find where the gradient of the tangent to the curve $y = 7 - 2x - 3x^2$ is horizontal

For example:
Find the coordinates of the points where the gradient of the tangent to the curve $y = 2x^3 - 9x^2 + 10x - 2$ is 34.

$$\frac{dy}{dx} = 6x^2 - 18x + 10$$

$$6x^2 - 18x + 10 = 34$$
$$6x^2 - 18x - 24 = 0$$

Dividing through by 6 gives:
$$x^2 - 3x - 4 = 0$$

This is a quadratic equation and you can solve it by factorising to get:
$$(x - 4)(x + 1) = 0$$

So $\qquad x = 4$ or – 1

Now we find y by substituting into the original equation:
$$y = 2x^3 - 9x^2 + 10x - 2$$

$x = 4$ gives:
$$y = 2(4)^3 - 9(4)^2 + 10(4) - 2$$
$$y = 128 - 144 + 40 - 2$$
$$y = 22$$

$x = -1$ gives:
$$y = 2(-1)^3 - 9(-1)^2 + 10(-1) - 2$$
$$y = -2 - 9 - 10 - 2$$
$$y = -23$$

Answers: $(4, 22)$ and $(-1, -23)$

Exercise 7C:

1. Find the coordinates of the points where the gradient of the tangent to the curve $y = 2x^3 - 3x^2 + 4x - 2$ is 4

2. Find the coordinates of the points where the gradient of the tangent to the curve $y = 5 + 2x - x^2 + 4x^3$ is 46

3. Find the coordinates of the points where the gradient of the tangent to the curve $y = 4x - 2x^2 + 3x^3$ is 17

4. Find the coordinates of the points where the gradient of the tangent to the curve $y = 3 + 2x - x^2 - 2x^3$ is – 18

5. Find the coordinates of the points where the gradient of the tangent to the curve $y = x^3 - 2x^2 + 5x - 2$ is 9

6. Find the coordinates of the points where the gradient of the tangent to the curve $y = \frac{1}{3}x^3 + \frac{3}{2}x^2 - 3x + 1$ is 1

7. Find the coordinates of the points where the gradient of the tangent to the curve $y = x^3 - x^2 + 4x - 3$ is 9

8. Find the coordinates of the points where the gradient of the tangent to the curve $y = \frac{2}{3}x^3 - \frac{3}{2}x^2 + x - 4$ is 3

9. Find the coordinates of the points where the gradient of the tangent to the curve $y = 4x^3 - 2x^2 + 5x + 1$ is 21

10. Find the coordinates of the points where the gradient of the tangent to the curve $y = 7 - 2x + 3x^2 - x^3$ is – 2

Exercise 7C...

11. Find the coordinates of the points where the gradient of the tangent to the curve $y = 4x^3 + 15x^2 - 18x + 2$ is is parallel to the x axis

12. Find the coordinates of the points where the gradient of the tangent to the curve $y = x^3 - 5x^2 - 8x - 3$ is horizontal

13. Find the coordinates of the points where the gradient of the tangent to the curve $y = x^3 - 3x^2 - 24x + 5$ is horizontal

7.2 Equation Of A Tangent

To find the equation of a tangent to a curve at any point you should:

- Find the gradient as before
- Substitute the gradient and coordinates of the point into $y = mx + c$

For example:
Find the equation of the tangent to the curve $y = 4x^3 + 15x^2 - 18x - 6$ at the point where $x = -3$

You must first find the y–coordinate of the point by substituting $x = -3$ into y to get:

$$y = 4(-3)^3 + 15(-3)^2 - 9(-3) - 6$$
$$y = -108 + 135 + 27 - 6$$
$$y = 48$$

Then differentiate the equation of the curve:

$$\frac{dy}{dx} = 12x^2 + 30x - 9$$

Substituting $x = -3$ gives:

$$\frac{dy}{dx} = 108 - 90 - 9$$

So $\quad m = 9$

Substituting $x = -3$, $y = 48$ and $m = 9$ into $y = mx + c$ gives:

$$y = 9x + c$$
$$48 = 9(-3) + c$$
$$48 = -27 + c$$
$$75 = c$$

Answer $\quad y = 9x + 75$

Exercise 7D:

1. Find the equation of the tangent to the curve $y = 2x^3 - 5x^2 + 3x - 2$ at the point where $x = 2$

2. Find the equation of the tangent to the curve $y = 4x^3 + x^2 - 2x + 1$ at $(1, 4)$

3. Find the equation of the tangent to the curve $y = 4 - 2x + 3x^2 - x^3$ at the point where $x = 3$

4. Find the equation of the tangent to the curve $y = 6 + 3x - 2x^2 + x^3$ at $(-2, -16)$

5. Find the equation of the tangent to the curve $y = 3x^3 - 4x^2 + 2x - 1$ at $(-1, -10)$

6. Find the equation of the tangent to the curve $y = x^3 - 2x^2 + 5x - 4$ at the point where $x = -2$

7. Find the equation of the tangent to the curve $y = \frac{2}{x^2}$ at $(-1, 2)$

8. Find the equation of the tangent to the curve $y = \frac{1}{3}x^3 - \frac{3}{2}x^2 + 4x - 1$ at the point where $x = 3$

9. Find the equation of the tangent to the curve $y = \frac{4}{x}$ where $x = 4$

10. Find the equation of the tangent to the curve $y = 2x + \frac{1}{2x}$ where $x = 1$

7.3 Equation of a Normal

A normal is a line which is perpendicular to a tangent at a point.
Remember that the product of two perpendicular gradients is -1.
To find the equation of a normal to a curve at any point you should:

- Find the gradient of the tangent as before.
- Find the gradient of the normal.
- Substitute this gradient and coordinates of the point into $y = mx + c$

For example:
1. Find the equation of the normal to the curve $y = x^3 - x^2 - 5x$ at $(2, -6)$

$$\frac{dy}{dx} = 3x^2 - 2x - 5$$

Substituting $x = 2$ gives:
$$12 - 4 - 5 = 3$$
So the gradient of the tangent = 3

So the gradient of the normal will be $\dfrac{-1}{3}$

Substituting $x = 2$, $y = -6$ and $m = \dfrac{-1}{3}$ into $y = mx + c$ gives:

$$y = \dfrac{-1}{3}x + c$$

$$-6 = \dfrac{-1}{3}(2) + c$$

$$-6 = \dfrac{-2}{3} + c$$

$$-6 + \dfrac{2}{3} = c$$

$$-5\dfrac{1}{3} = c$$

Answer $\qquad y = \dfrac{-1}{3}x - 5\dfrac{1}{3}$

Exercise 7E:

1. Find the equation of the normal to the curve $y = 2x^3 - 3x^2 + 5x$ at $(2, 14)$

2. Find the equation of the normal to the curve $y = 3x^2 + x - 4$ where $x = 1$

3. Find the equation of the normal to the curve $y = x^3 - 2x^2 + x - 5$ at $(-2, -23)$

4. Find the equation of the normal to the curve $y = 4x^3 - x^2 + 5x - 1$ where $x = -1$

5. Find the equation of the normal to the curve $y = 7 - 2x + 3x^2 - x^3$ at the point where $x = 2$

6. Find the equation of the normal to the curve $y = 10 + 3x - x^2 - 4x^3$ at $(1, 8)$

7. Find the equation of the normal to the curve $y = \dfrac{3}{x^2}$ where $x = 2$

8. Find the equation of the normal to the curve $y = \dfrac{4}{3}x^3 - \dfrac{1}{2}x^2 + x - 2$ at $(-1, \dfrac{-29}{6})$

9. Find the equation of the normal to the curve $y = \dfrac{2}{x}$ where $x = -2$

10. Find the equation of the normal to the curve $y = 4x^2 - \dfrac{1}{4x}$ where $x = -1$

Further examples:

1. (a) Find the equation of the tangent to the curve
$$y = 2x^2 - 4x - 15 \text{ at } (-1, -9)$$
   (b) Find where the tangent crosses the
$$\text{(i) } y\text{-axis (ii) } x\text{-axis}$$

(a) $\qquad \dfrac{dy}{dx} = 4x - 4$

Substituting $x = -1$ gives:

$$\dfrac{dy}{dx} = -4 - 4$$
$$= -8$$

So $\qquad m = -8$

Substituting $x = -1$, $y = -9$ and $m = -8$ into $y = mx + c$ gives:

$$y = -8x + c$$
$$-9 = -8(-1) + c$$
$$-9 = 8 + c$$
$$-17 = c$$

Answer $\qquad y = -8x - 17$

(b)(i) The tangent will cross the y-axis when $x = 0$
So, substituting $x = 0$ into $y = -8x - 17$ gives:

$$y = -17$$

Answer $\qquad (0, -17)$

(ii) The tangent will cross the x-axis when $y = 0$
So, substituting $y = 0$ into $y = -8x - 17$ gives:

$$0 = -8x - 17$$
$$8x = -17$$
$$x = \dfrac{-17}{8}$$

Answer $\qquad (\dfrac{-17}{8}, 0)$

2. (a) Find the equation of the normal to the curve
$$y = 7 - 4x - x^2 \text{ at } (2, -5)$$
   (b) Find where the normal meets the straight line
$$8y + 3x = 2$$

(a) $\qquad \dfrac{dy}{dx} = -4 - 2x$

Substituting $x = 2$ gives:
$$\dfrac{dy}{dx} = -4 - 4$$
$$= -8$$

So the gradient of the tangent $= -8$

Therefore the gradient of the normal will be $\dfrac{1}{8}$

Substituting $x = 2$, $y = -5$ and $m = \dfrac{1}{8}$ into $y = mx + c$ gives:

$$y = \frac{1}{8}x + c$$

$$-5 = \frac{1}{8}(2) + c$$

$$-5 - \frac{1}{4} = c$$

$$-5\frac{1}{4} = c$$

Answer $\quad y = \frac{1}{8}x - 5\frac{1}{4}$

(b) You can find where the normal $y = \frac{1}{8}x - 5\frac{1}{4}$ meets the line $8y + 3x = 2$ by solving these simultaneous equations.

$$\left[y = \frac{1}{8}x - 5\frac{1}{4} \right] \times 8 \text{ gives:}$$

$$8y = x - 42$$

You can rewrite this as $8y - x = -42$ and then solve by subtraction:

$$
\begin{array}{r}
8y - x = -42 \\
- \underline{8y + 3x = 2} \\
-4x = -44 \\
x = 11
\end{array}
$$

You can substitute $x = 11$ into either equation to get y:

$$y = \frac{1}{8}x - 5\frac{1}{4}$$

$$y = \frac{1}{8}(11) - 5\frac{1}{4}$$

$$y = 1\frac{3}{8} - 5\frac{1}{4}$$

$$y = -3\frac{7}{8}$$

Answer $\quad (11, -3\frac{7}{8})$

Exercise 7F:

1. (a) Find the equation of the tangent to the curve $y = 4x^2 - 3x + 5$ at $(2, 15)$

   (b) Find where the tangent crosses the
   (i) y–axis (ii) x–axis

2. (a) Find the equation of the normal to the curve $y = 4 + 3x + 2x^2$ at $(-3, 13)$

   (b) Find where the normal crosses the
   (i) y–axis (ii) x–axis

3. (a) Find the equation of the tangent to the curve $y = x^3 + 2x^2 - 5x + 3$ at $(-1, 9)$

Exercise 7F...

   (b) Find where the tangent meets the straight line $y + 4x = 9$

4. (a) Find the equation of the normal to the curve $y = 7 - x + x^2 - x^3$ at $(2, 1)$

   (b) Find where the normal meets the straight line $3y + x = 9$

5. (a) Find the equation of the tangent to the curve $y = \frac{2}{3}x^3 - \frac{1}{2}x^2 + x$ where $x = -1$

   (b) Find where the tangent crosses the y–axis

6. (a) Find the equation of the normal to the curve $y = (3x + 2)^2$ at $(-1, 1)$

   (b)(i) Find where the normal crosses the x–axis

   (ii) Find where the normal meets the straight line $4y + x = 8$

7. (a) Find the equation of the tangent to the curve $y = 2x^3 + 3x^2 - x + 5$ at $(-2, 3)$

   (b)(i) Find where the tangent crosses the x–axis

   (ii) Find where the tangent meets the straight line $y = 3x + 1$

8. (a) Find the equation of the normal to the curve $y = 5x - x^2 + 4$ at $(-1, -2)$

   (b)(i) Find where the normal crosses the y–axis

   (ii) Find where the normal meets the straight line $y = x - 9$

9. (a) Find the equation of the tangent to the curve $y = (x - 4)(2x - 1)$ at $(2, -6)$

   (b)(i) Find where the tangent crosses the y–axis

   (ii) Find where the tangent meets the straight line $y + 2x = 1$

10. (a) Find the equation of the normal to the curve $y = \frac{5}{2x^2}$ where $x = -1$

    (b)(i) Find where the normal crosses the x–axis

    (ii) Find where the normal meets the straight line $2y + x = 7$

You can use a similar technique to find where a tangent is parallel to a straight line.

The rules are as follows:

- Find the gradient of the tangent (by differentiating).
- Find the gradient of the straight line (rearranging equation into $y = mx + c$).
- Put the two expressions equal to each other, and solve for x.
- Substitute the value for x into the equation of the curve to find the y coordinate.

For example:

1. Find where the tangent to the curve $y = 5x^2 - 6x + 1$ is parallel to the straight line $x + y = 4$

The gradient of the tangent is:
$$\frac{dy}{dx} = 10x - 6$$

Rearranging the equation of the straight line gives:
$$x + y = 4$$
$$y = -x + 4$$

So the gradient of the straight line is -1

The gradients are equal, so:
$$10x - 6 = -1$$
$$10x = -1 + 6$$
$$10x = 5$$
$$x = \frac{5}{10} = \frac{1}{2}$$

Substituting the value for x into the equation of the curve gives:
$$y = 5x^2 - 6x + 1$$
$$y = 5(\tfrac{1}{2})^2 - 6(\tfrac{1}{2}) + 1$$
$$y = \frac{5}{4} - 3 + 1$$
$$y = \frac{-3}{4}$$

Answer $\quad (\frac{1}{2}, \frac{-3}{4})$

....

2. Find where the tangent to the curve $y = 5x - x^2 + 4$ is perpendicular to the straight line joining the points $(-2, 3)$ and $(3, 2)$.

In this case, we need to:

- Find the gradient of the tangent (by differentiating).
- Find the gradient of the straight line (using the coordinate formula).
- Find the gradient of the perpendicular to the straight line.

The gradient of the tangent is:
$$\frac{dy}{dx} = 5 - 2x$$

The gradient of a straight line is given by $\frac{y_2 - y_1}{x_2 - x_1}$

$$\text{Gradient} = \frac{2 - 3}{3 - (-2)}$$
$$= \frac{-1}{5}$$

So the gradient of the perpendicular $= 5$

Putting these equal gives:
$$5 - 2x = 5$$
$$-2x = 0$$
$$x = 0$$

Substituting the value for x into the equation of the curve gives:
$$y = 5x - x^2 + 4$$
$$y = 5(0) - (0)^2 + 4$$
$$y = 4$$

Answer $= (0, 4)$

3. Find the coordinates of the points where the tangent to the curve $y = 4x^3 - 15x^2 - 60x + 2$ is parallel to the straight line $y = 4 + 12x$

The gradient of the tangent is:
$$\frac{dy}{dx} = 12x^2 - 30x - 60$$

Rearranging the equation of the straight line gives:
$$y = 4 + 12x$$
$$y = 12x + 4$$

So the gradient of the straight line is 12

Putting these equal gives:
$$12x^2 - 30x - 60 = 12$$
$$12x^2 - 30x - 72 = 0$$

Dividing by 6 gives:
$$2x^2 - 5x - 12 = 0$$
$$(2x + 3)(x - 4) = 0$$

So $\quad x = \frac{-3}{2}$ or 4

Substituting into the equation of the curve...

where $x = \frac{-3}{2}$:
$$y = 4x^3 - 15x^2 - 60x + 2$$
$$y = 4(\frac{-3}{2})^3 - 15(\frac{-3}{2})^2 - 60(\frac{-3}{2}) + 2$$
$$y = -13\frac{1}{2} - 33\frac{3}{4} + 90 + 2$$
$$y = 44\frac{3}{4}$$

where $x = 4$:

$$y = 4x^3 - 15x^2 - 60x + 2$$
$$y = 4(4)^3 - 15(4)^2 - 60(4) + 2$$
$$y = -222$$

Answer $(\frac{-3}{2}, 44\frac{3}{4})$ and $(4, -222)$

Exercise 7G:

1. Find where the tangent to the curve
   $y = 2x^2 - 3x + 2$ is parallel to the straight line
   $y = 5x - 1$

2. Find where the tangent to the curve
   $y = 7 + 2x - 3x^2$ is perpendicular to the straight
   line $4y - x = 5$

3. Find where the tangent to the curve
   $y = x^3 - 2x^2 + x - 4$ is parallel to the x axis

4. Find where the tangent to the curve
   $y = 7 - 2x + 2\frac{1}{2}x^2 - 2x^3$ is parallel to the
   straight line $y + 3x = 4$

5. Find where the tangent to the curve $y = \frac{4}{x}$ is
   perpendicular to the straight line $9y = 6 + 4x$

Exercise 7G...

6. Find where the tangent to the curve
   $y = 4 - 6x + x^2$ is perpendicular to the straight
   line $2y + x = 5$

7. Find where the tangent to the curve
   $y = (x - 3)(3x + 1)$ is parallel to the straight line
   $y = 5 + x$

8. Find where the tangent to the curve $y = 4x + \frac{4}{x}$
   is perpendicular to the straight line $5y = x - 2$

9. Find where the tangent to the curve
   $y = x^3 - 4x^2 + 2x - 1$ is parallel to the straight
   line $y + 2x = 1$

10. Find where the tangent to the curve
    $y = \frac{2}{3}x^3 - \frac{9}{2}x^2 - 3x + 1$ is perpendicular to the
    straight line $4y + 2x = 7$

CHAPTER 8: TURNING POINTS

8.1 Types of Turning Point

A curve can have two types of turning point. A turning
point is where the gradient of a curve changes from either

(a) positive to 0 to negative, as shown below.

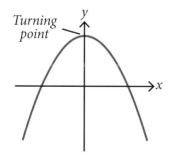

This is called a **maximum** turning point.

(b) negative to 0 to positive, as shown below:

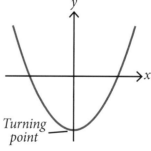

This is called a **minimum** turning point.

The **gradient of the curve is 0** at both types of turning
point.

You can therefore find the turning point(s) of a curve by following these rules:

- Differentiate to find $\frac{dy}{dx}$
- Put $\frac{dy}{dx} = 0$ and solve to find x.
- Substitute x into y to find the y coordinate if necessary.

You need to be able to distinguish between maximum and minimum turning points. You can do this either by finding the gradients before and after the turning point, or by differentiating $\frac{dy}{dx}$ again to get $\frac{d^2y}{dx^2}$.

$\frac{d^2y}{dx^2}$ measures the **rate** of change of the gradient $\frac{dy}{dx}$ as x changes.

For a maximum turning point the gradient changes from positive to 0 to negative, ie it is getting smaller.

So if $\frac{d^2y}{dx^2}$ is negative the turning point will be a maximum.

For a minimum turning point the gradient changes from negative to 0 to positive, ie it is getting bigger.

So if $\frac{d^2y}{dx^2}$ is positive the turning point will be a minimum.

When you are asked to say what type of turning point there is, you must justify your answer by using either of the above methods.

8.2 Quadratic Curves

For example, find the turning points of the following curves. Say what type each turning point is.

1. $y = 2x^2 - 7x + 3$

Differentiate: $\frac{dy}{dx} = 4x - 7$

Now set this equal to 0 to find the x coordinate of the turning point:
$$4x - 7 = 0$$
$$4x = 7$$
$$x = 1.75$$

Substitute this back into the original equation to find the y coordinate:
$$y = 2(1.75)^2 - 7(1.75) + 3$$
$$y = -3.125$$

The turning point is at $(1.75, -3.125)$

You must now say what type it is and justify this. Either:

- Substitute $x = 1$ into $\frac{dy}{dx}$ to get $4 - 7 = -3$

And $x = 2$ into $\frac{dy}{dx}$ to get $8 - 7 = 1$

It must be a minimum because the gradient is getting bigger.

- Or find $\frac{d^2y}{dx^2} = 4$

It must be a minimum because $\frac{d^2y}{dx^2}$ is positive.

....

2. $y = 7 - 8x - x^2$

Differentiate: $\frac{dy}{dx} = -8 - 2x$

Now set this equal to 0 to find the x coordinate of the turning point:
$$-8 - 2x = 0$$
$$-2x = 8$$
$$x = -4$$

Substitute this back into the original equation to find the y coordinate:
$$y = 7 - 8(-4) - (-4)^2$$
$$y = 23$$

The turning point is at $(-4, 23)$

You must now say what type it is and justify this. Either:

- Substitute $x = -5$ into $\frac{dy}{dx}$ to get $-8 + 10 = 2$

And $x = 0$ into $\frac{dy}{dx}$ to get $-8 - 0 = -8$

It must be a maximum because the gradient is getting smaller.

- Or find $\frac{d^2y}{dx^2} = -2$

It must be a maximum because $\frac{d^2y}{dx^2}$ is negative.

....

3. $y = (x + 4)(2x - 1)$

You must first work out the brackets to get:
$$y = x(2x - 1) + 4(2x - 1)$$
$$y = 2x^2 - 1x + 8x - 4$$
$$y = 2x^2 + 7x - 4$$

Differentiate: $\frac{dy}{dx} = 4x + 7$

Now set this equal to 0 to find the x coordinate of the turning point:
$$4x + 7 = 0$$
$$4x = -7$$
$$x = -1.75$$

Substitute this back into the original equation to find the y coordinate:
$$y = 2(-1.75)^2 + 7(-1.75) - 4$$
$$y = -10.125$$

The turning point is at $(-1.75, -10.125)$

You must now say what type it is and justify this. Either:

- Substitute $x = -2$ into $\frac{dy}{dx}$ to get $-8 + 7 = -1$

  And $x = 0$ into $\frac{dy}{dx}$ to get $0 + 7 = 7$

It must be a minimum because the gradient is getting bigger.

- Or find $\frac{d^2y}{dx^2} = 4$

  It must be a minimum because $\frac{d^2y}{dx^2}$ is positive.

Exercise 8A: *Find the turning points of the following curves. Say what type each turning point is.*

1. $y = 3x^2 - 12x + 1$	**6.** $y = 1 + 7x - x^2$
2. $y = x^2 + 4x - 7$	**7.** $y = 4x^2 + 2x - 5$
3. $y = 9 + 6x - x^2$	**8.** $y = 5 - x - x^2$
4. $y = x^2 - 5x - 3$	**9.** $y = x^2 - 6x - 7$
5. $y = 3 - 8x - 2x^2$	**10.** $y = 8 - 4x - x^2$

8.3 Cubic Curves

A cubic curve has two turning points: a maximum and a minimum.

For example, find the turning points of the following curves. Say what type each turning point is.

1. $y = 4x^3 + 2x^2 - 6x + 4$

Differentiate: $\frac{dy}{dx} = 12x^2 + 4x - 6$

Now set this equal to 0 to find the x coordinates of the turning points:

$$12x^2 + 4x - 6 = 0$$

Solve using the quadratic formula $\frac{-b \pm \sqrt{b^2 - 4ac}}{2a}$:

Where $a = 12, b = 4, c = -6$

So $x = \dfrac{-4 \pm \sqrt{4^2 - (4 \times 12 \times -6)}}{2 \times 12}$

$x = \dfrac{-4 \pm \sqrt{304}}{24}$

$x = \dfrac{-4 + \sqrt{304}}{24}$ or $\dfrac{-4 - \sqrt{304}}{24}$

So $x = 0.560$ or -0.893 to 3 decimal places.

Substitute these back into the original equation to find the y coordinates:

$x = 0.560 \quad y = 4(0.560)^3 + 2(0.560)^2 - 6(0.560) + 4$
$\qquad\qquad y = 1.970$

$x = -0.893 \quad y = 4(-0.893)^3 + 2(-0.893)^2 - 6(-0.893) + 4$
$\qquad\qquad y = 8.104$

So the turning points are at (0.560, 1.970) and (−0.893, 8.104).

Now show what type each turning point is by getting $\frac{d^2y}{dx^2}$.

$$\frac{dy}{dx} = 12x^2 + 4x - 6$$

$$\frac{d^2y}{dx^2} = 24x + 4$$

When $x = 0.560$ $\frac{d^2y}{dx^2} = 24(0.560) + 4 = 17.44$ and so it is a minimum.

When $x = -0.893$ $\frac{d^2y}{dx^2} = 24(-0.893) + 4 = -17.432$ and so it is a maximum.

2. $y = 3 + 2x - x^2 - 2x^3$

Rearrange: $y = -2x^3 - x^2 + 2x + 3$

Differentiate: $\frac{dy}{dx} = -6x^2 - 2x + 2$

Now set this equal to 0 to find the x coordinates of the turning points:

$$-6x^2 - 2x + 2 = 0$$

Solve using the quadratic formula $\frac{-b \pm \sqrt{b^2 - 4ac}}{2a}$:

Where $a = -6, b = -2, c = 2$

So $x = \dfrac{2 \pm \sqrt{(-2)^2 - (4 \times -6 \times 2)}}{2 \times -6}$

$x = \dfrac{2 \pm \sqrt{52}}{-12}$

$x = \dfrac{2 + \sqrt{52}}{-12}$ or $\dfrac{2 - \sqrt{52}}{-12}$

So $x = -0.768$ or 0.434 to 3 decimal places.

Substitute these back into the original equation to find the y coordinates:

$x = -0.768 \quad y = 3 + 2x - x^2 - 2x^3$
$\qquad\qquad y = 3 + 2(-0.768) - (-0.768)^2 - 2(-0.768)^3$
$\qquad\qquad y = 1.780$

$x = 0.434 \quad y = 3 + 2(0.434) - (0.434)^2 - 2(0.434)^3$
$\qquad\qquad y = 3.516$

So the turning points are at (−0.768, 1.780) and (0.434, 3.516).

Now show what type each turning point is by getting $\frac{d^2y}{dx^2}$.

$$\frac{dy}{dx} = -6x^2 - 2x + 2$$

$$\frac{d^2y}{dx^2} = -12x - 2$$

When $x = -0.768$ $\frac{d^2y}{dx^2} = -12(-0.768) - 2 = 7.216$ and so it is a minimum.

When $x = 0.434$ $\frac{d^2y}{dx^2} = -12(0.434) - 2 = 7.208$ and so it is a maximum.

3. $y = x(x + 4)(2x - 1)$

First expand the brackets to get:
$$y = 2x^3 + 7x^2 - 4x$$

Differentiate: $\frac{dy}{dx} = 6x^2 + 14x - 4$

Now set this equal to 0 to find the x coordinates of the turning points:
$$6x^2 + 14x - 4 = 0$$

Solve using the quadratic formula $\frac{-b \pm \sqrt{b^2 - 4ac}}{2a}$:

Where $\qquad a = 6, b = 14, c = -4$

So $\qquad\qquad x = \dfrac{-14 \pm \sqrt{(14)^2 - (4 \times 6 \times -4)}}{2 \times 6}$

$\qquad\qquad\qquad x = \dfrac{-14 \pm \sqrt{292}}{12}$

$\qquad\qquad\qquad x = \dfrac{-14 + \sqrt{292}}{12}$ or $\dfrac{-14 - \sqrt{292}}{12}$

So $\qquad\qquad x = 0.257$ or -2.591 to 3 decimal places.

Substitute these back into the original equation to find the y coordinates:

$x = 0.257 \qquad y = x(x + 4)(2x - 1)$
$\qquad\qquad\quad y = 0.257(0.257 + 4)(2 \times 0.257 - 1)$
$\qquad\qquad\quad y = -0.532$

$x = -2.591 \qquad y = -2.591(-2.591 + 4)(2 \times -2.591 - 1)$
$\qquad\qquad\quad y = 22.569$

So the turning points are at $(0.257, -0.532)$ and $(-2.591, 22.569)$.

Now show what type each turning point is by getting $\frac{d^2y}{dx^2}$.
$$\frac{dy}{dx} = 6x^2 + 14x - 4$$

$$\frac{d^2y}{dx^2} = 12x + 14$$

When $x = 0.257$ $\frac{d^2y}{dx^2} = 12(0.257) + 14 = 17.084$ and so it is a minimum.

When $x = -2.591$ $\frac{d^2y}{dx^2} = 12(-2.591) + 14 = -17.092$ and so it is a maximum.

Exercise 8B: *Find the turning points of the following curves. Say what type each turning point is.*

1. $y = 2x^3 + 3x^2 - 36x + 4$
2. $y = 2x^3 - 9x^2 - 24x + 3$
3. $y = x^3 - 9x^2 + 24x - 7$
4. $y = x^3 + 4x^2 - 3x - 2$
5. $y = 4x^3 - 3x + 2$
6. $y = x^3 - 5x^2 + 8x + 2$
7. $y = 4x^3 - 23x^2 + 30x - 5$

8. $y = x^3 + 5x^2 - 8x$
9. $y = 4x^3 - x^2 - 30x + 3$
10. $y = x^3 + 3x^2 - 72x - 8$
11. $y = 5 + 45x - 3x^2 - x^3$
12. $y = 3 - 12x - 9x^2 - 2x^3$
13. $y = 24 - 12x + 15x^2 - 4x^3$

8.4 Curve Sketching – Quadratic Curves

To sketch a quadratic curve, you need to find three points:
- Find where the curve crosses the y axis by putting $x = 0$;
- Find where the curve crosses the x axis by putting $y = 0$;
- Find the turning point.

Then mark these points on a grid, and draw a smooth curve through these points.

For example, sketch the following:

1. $y = 3x^2 + 22x - 16$

The curve crosses the y axis when $x = 0$
So $\qquad\qquad y = 0 + 0 - 16$
$\qquad\qquad\quad = -16$

The curve crosses the x axis when $y = 0$
So $3x^2 + 22x - 16 = 0$
$\qquad (3x - 2)(x + 8) = 0$
Either $\qquad 3x - 2 = 0$
$\qquad\qquad\quad 3x = 2$
$\qquad\qquad\quad x = \dfrac{2}{3}$

or $\qquad\qquad x + 8 = 0$
$\qquad\qquad\quad x = -8$

The turning point is when $\frac{dy}{dx} = 0$
$$y = 3x^2 + 22x - 16$$

$$\frac{dy}{dx} = 6x + 22 = 0$$
$$6x = -22$$
$$x = \frac{-22}{6} = \frac{-11}{3} \text{ (or } -3\tfrac{2}{3})$$

You should work with fractions or mixed numbers rather than decimals.

You now find the y coordinate of the turning point by substituting $x = \frac{-11}{3}$ into the equation for y
$$y = 3x^2 + 22x - 16$$

$$y = 3(\tfrac{-11}{3})^2 + 22(\tfrac{-11}{3}) - 16$$

$$y = \frac{121}{3} - \frac{242}{3} - 16 = -56\tfrac{1}{3}$$

The turning point is at $(-3\tfrac{2}{3}, -56\tfrac{1}{3})$

Now show what type of turning point it is by getting $\frac{d^2y}{dx^2}$.
$$\frac{dy}{dx} = 6x + 22$$
$$\frac{d^2y}{dx^2} = 6$$

So it is a minimum.
You can now sketch the curve:

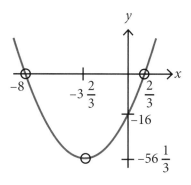

1. $y = x^2 - 6x + 8$ **6.** $y = 3x^2 - 12x$

2. $y = 5 - 4x - x^2$ **7.** $y = 2 - x - 6x^2$

3. $y = 2x^2 + 5x - 3$ **8.** $y = x^2 + 9x + 18$

4. $y = 3x^2 - 10x - 8$ **9.** $y = 8x - 12x^2$

5. $y = 1 - 16x^2$ **10.** $y = 10 - 13x - 3x^2$

2. $y = 9 - 4x^2$

The curve crosses the y axis when $x = 0$

So $\qquad\qquad y = 9 - 0$

$\qquad\qquad\quad\ = 9$

The curve crosses the x axis when $y = 0$

So $\qquad 9 - 4x^2 = 0$

$\quad (3 - 2x)(3 + 2x) = 0$

Either $\qquad 3 - 2x = 0$

$\qquad\qquad\ \ -2x = -3$

$\qquad\qquad\qquad x = \dfrac{3}{2}$

or $\qquad\quad 3 + 2x = 0$

$\qquad\qquad\quad 2x = -3$

$\qquad\qquad\qquad x = -\dfrac{3}{2}$

The turning point is when $\dfrac{dy}{dx} = 0$

$\qquad\qquad\quad y = 9 - 4x^2$

$\qquad\qquad \dfrac{dy}{dx} = -8x = 0$

$\qquad\qquad\qquad x = 0$

You already know the value of y when $x = 0$ (ie 9)

So the turning point is at (0, 9)

Now show what type of turning point it is by getting $\dfrac{d^2y}{dx^2}$

$\qquad\qquad \dfrac{dy}{dx} = -8x$

$\qquad\qquad \dfrac{d^2y}{dx^2} = -8$

So it is a maximum.

You can now sketch the curve :

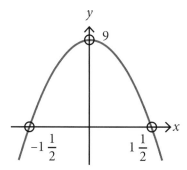

8.5 Curve Sketching – Cubic Curves

Any cubic functions you would get in your exam will always have x as a common factor. This means any cubic function that you are asked to sketch will go through the origin. The rules are as follows:

• Find where the curve crosses the x axis by putting $y = 0$ and taking out the common factor. Remember that (0, 0) will always be an answer.

• Find the turning points.

• Distinguish between the turning points, clearly showing your method.

• Mark these points on a grid.

• Draw a smooth curve through these points.

For example:

1. Sketch the curve $y = 4x^3 + 7x^2 - 10x$ giving the coordinates of the points where it crosses the x axis correct to 2 decimal places where appropriate.

The curve crosses the x axis when $y = 0$

So $4x^3 + 7x^2 - 10x = 0$

$\quad x(4x^2 + 7x - 10) = 0$

So either $x = 0$ or $4x^2 + 7x - 10 = 0$

You need to use the quadratic formula $\dfrac{-b \pm \sqrt{b^2 - 4ac}}{2a}$ to solve $4x^2 + 7x - 10 = 0$

Where $\qquad\qquad a = 4, b = 7, c = -10$

So $\qquad\qquad\quad x = \dfrac{-7 \pm \sqrt{(7)^2 - (4 \times 4 \times -10)}}{2 \times 4}$

$\qquad\qquad\qquad x = \dfrac{-7 \pm \sqrt{209}}{8}$

$\qquad\qquad\qquad x = \dfrac{-7 + \sqrt{209}}{8}$ or $\dfrac{-7 - \sqrt{209}}{8}$

$\qquad\qquad\qquad x = 0.93$ or -2.68, to 2 decimal places

The curve crosses the x axis at (0, 0) (0.93, 0) and (−2.68, 0)

The turning points are when $\frac{dy}{dx} = 0$

$$y = 4x^3 + 7x^2 - 10x$$

$$\frac{dy}{dx} = 12x^2 + 14x - 10 = 0$$

Dividing through by 2 gives:

$$6x^2 + 7x - 5 = 0$$
$$(2x - 1)(3x + 5) = 0$$

So either $2x - 1 = 0$
$$2x = 1$$
$$x = \frac{1}{2}$$

or $3x + 5 = 0$
$$3x = -5$$
$$x = \frac{-5}{3}$$

You now find y by substituting $x = \frac{1}{2}$ or $\frac{-5}{3}$ into the equation for y

$$y = 4x^3 + 7x^2 - 10x$$

$$y = 4(\tfrac{1}{2})^3 + 7(\tfrac{1}{2})^2 - 10(\tfrac{1}{2})$$

$$y = \frac{1}{2} + \frac{7}{4} - 5 = -2\frac{3}{4}$$

This turning point is at $(\frac{1}{2}, -2\frac{3}{4})$

$$y = 4x^3 + 7x^2 - 10x$$

$$y = 4(\tfrac{-5}{3})^3 + 7(\tfrac{-5}{3})^2 - 10(\tfrac{-5}{3})$$

$$y = \frac{-500}{27} + \frac{175}{9} + \frac{50}{3} = 17\frac{16}{27}$$

This turning point is at $(\frac{-5}{3}, 17\frac{16}{27})$

Now show what type each turning point is by getting $\frac{d^2y}{dx^2}$

$$\frac{dy}{dx} = 12x^2 + 14x - 10$$

$$\frac{d^2y}{dx^2} = 24x + 14$$

When $x = \frac{1}{2}$: $\frac{d^2y}{dx^2} = 12 + 14 = 26$ and so it is a minimum.

When $x = \frac{-5}{3}$: $\frac{d^2y}{dx^2} = -40 + 14 = -26$ and so it is a maximum.

You can now sketch the curve:

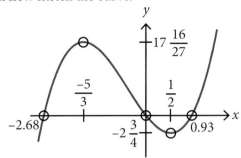

2. Sketch the curve $y = x^3 - 9x^2 + 24x$

The curve crosses the x axis when $y = 0$

So $x^3 - 9x^2 + 24x = 0$
$$x(x^2 - 9x + 24) = 0$$

So either $x = 0$ or $x^2 - 9x + 24 = 0$

You need to use the quadratic formula $\frac{-b \pm \sqrt{b^2 - 4ac}}{2a}$ to solve $x^2 - 9x + 24 = 0$

Where $a = 1, b = -9, c = 24$

So $x = \dfrac{9 \pm \sqrt{(-9)^2 - (4 \times 1 \times 24)}}{2 \times 1}$

$$x = \frac{9 \pm \sqrt{-15}}{2}$$

But $\sqrt{-15}$ does not exist.

Therefore this curve only crosses the x axis at $(0, 0)$.

The turning points are when $\frac{dy}{dx} = 0$

$$y = x^3 - 9x^2 + 24x$$

$$\frac{dy}{dx} = 3x^2 - 18x + 24 = 0$$

Dividing through by 3 gives:

$$x^2 - 6x + 8 = 0$$
$$(x - 2)(x - 4) = 0$$

So either $x - 2 = 0$
$$x = 2$$

or $x - 4 = 0$
$$x = 4$$

You now find y by substituting $x = 2$ or 4 into the equation for y

$$y = x^3 - 9x^2 + 24x$$
$$y = (2)^3 - 9(2)^2 + 24(2)$$
$$y = 8 - 36 + 48 = 20$$

This turning point is at $(2, 20)$.

$$y = x^3 - 9x^2 + 24x$$
$$y = (4)^3 - 9(4)^2 + 24(4)$$
$$y = 64 - 144 + 96 = 16$$

This turning point is at $(4, 16)$.

Now show what type each turning point is by getting $\frac{d^2y}{dx^2}$

$$\frac{dy}{dx} = 3x^2 - 18x + 24$$

$$\frac{d^2y}{dx^2} = 6x - 18$$

When $x = 2$: $\frac{d^2y}{dx^2} = 12 - 18 = -6$ and so it is a maximum.

When $x = 4$: $\frac{d^2y}{dx^2} = 24 - 18 = 6$ and so it is a minimum.

You can now sketch the curve:

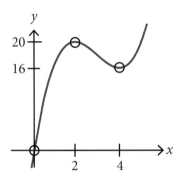

Exercise 8D: *Sketch the following curves, giving the coordinates of the point(s) where each crosses the x axis correct to 2 decimal places where appropriate:*

1. $y = x^3 - 3x^2 - 24x$

2. $y = 4x^3 - 21x^2 - 24x$

3. $y = 3x^3 - x$

4. $y = 2x^3 - 3x^2 - 120x$

5. $y = x^3 + 6x^2 - 36x$

6. $y = x^3 - 2x^2$

7. $y = 2x^3 - 21x^2 + 60x$

8. $y = x(x - 3)(x + 8)$

9. $y = 2x(2x - 1)(x + 3)$

10. $y = 3x(x - 2)(2x + 5)$

8.6 Optimising Problems

You can use differentiation in solving optimising problems. For example:

1. The sum of two numbers is 12. Find the minimum value of the sum of their squares.

Call the numbers x and y. So:
$$x + y = 12$$
The sum of their squares, S, is given by:
$$S = x^2 + y^2$$
Then replace one of the variables in terms of the other.
$$y = 12 - x$$
This gives: $\quad S = x^2 + (12 - x)^2$
$$= x^2 + x^2 - 24x + 144$$
$$= 2x^2 - 24x + 144$$

Then find the minimum value by differentiating S with respect to x and equating this to 0:
$$S = 2x^2 - 24x + 144 = 0$$
$$\frac{dS}{dx} = 4x - 24$$
$$4x - 24 = 0$$
$$4x = 24$$
$$x = 6$$

You should prove that this gives a minimum value by finding $\frac{d^2S}{dx^2}$:

$$\frac{d^2S}{dx^2} = 4 \text{ and so it is a minimum}$$

You can now find the value of y:
$$y = 12 - x$$
So $\quad y = 12 - 6$
$$y = 6$$

You can now find the minimum value of the sum of their squares:
$$S = x^2 + y^2$$
$$S = 6^2 + 6^2$$
$$S = 36 + 36$$
$$S = 72$$

Exercise 8E:

1. The difference of two numbers x and y is 8 where x is greater than y.
Find the minimum value of $4x^2 - 3y^2$

2. A farmer wants to enclose a rectangular area using a hedge as one side as shown below.

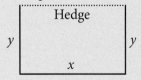

The perimeter has to be 60 m. Find:
(i) the values of x and y that give a maximum area
(ii) the maximum possible area.

3. ABC is a right-angled triangle in which angle A = 90°, AB = y cm, AC = x cm and BC = $2x$ cm.
The perimeter of ABC is 30 cm. Find:
(i) the values of x and y that give a maximum area
(ii) the exact maximum possible area.

4. x pencils cost y p each.
y pens cost $2x$ p each.
The total number of pencils and pens must be 20.
Find: (i) the values of x and y for which the total cost will be a maximum.
(ii) the maximum possible cost.

5. A farmer wants to enclose an area in the shape of a trapezium using a hedge as one side as shown below.

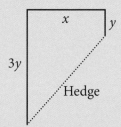

The area has to be 288 m². Find: (i) the values of x and y that give a minimum perimeter.
(ii) the minimum possible perimeter.

6. A farmer wants to enclose a rectangular area using a hedge as one side as shown below.

The area has to be 338 m². Find: (i) the values of x and y that give a minimum perimeter.
(ii) the minimum possible perimeter.

7. The perimeter of the L-shape below has to be 48 cm.

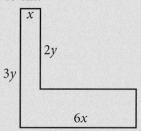

Find:
(i) the values of x and y that give a maximum area.
(ii) the maximum possible area.

8. The area of the trapezium below has to be 66 cm².

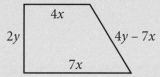

Find: (i) the values of x and y that give a minimum perimeter.
(ii) the minimum possible perimeter.

CHAPTER 9: INTEGRATION

9.1 Integration

Consider $\quad y = 4x^3 + 6x^2$

Then $\quad \dfrac{dy}{dx} = 12x^2 + 12x$

Suppose you started with the answer for $\dfrac{dy}{dx}$ and wanted to find the function y.

You would then have to do the exact opposite each time, ie:
• add 1 on to the power;
• divide by the new power.

Thus for $12x^2$ you would:
• add 1 on to the power to get 3
• divide 12 by 3 to get 4

Similarly for $12x$ you would:
• add 1 on to the power to get 2
• divide 12 by 2 to get 6

This process is called **integration**.

You write this as: $\int (12x^2 + 12x)\, dx$
The 'dx' shows that you are integrating with respect to x.

Now consider $\quad y = 4x + 6$

Then $\quad \dfrac{dy}{dx} = 4$

Again, suppose you started with the answer for $\dfrac{dy}{dx}$ and wanted to find the function y.

You would again have to do the exact opposite ie:
• add 1 on to the power;
• divide by the new power.

The power of x in the number 4 is 0
You would add 1 on to 0 to get 1
You would then divide 4 by 1 to get 4
You write this as: $\int 4\, dx$

The Constant Of Integration
In the second example, you may have noticed that we did not arrive back at the original equation by integrating.
Consider $\quad y = 4x + 7$

Then $\quad \dfrac{dy}{dx} = 4$
Now consider $\quad y = 4x - 60$

Then $\quad \dfrac{dy}{dx} = 4$

And consider $y = 4x$

Then $\dfrac{dy}{dx} = 4$

In each example you get 4 as the answer, even though the equations are different. This is because when you differentiate a constant number you get 0. This means that when you integrate **any** function you must always include **C**, which is known as the **constant of integration**. C can take any value. Thus the complete answers to the two integrations above are:

1. $\int (12x^2 + 12x)\, dx = 4x^3 + 6x^2 + C$
2. $\int 4\, dx = 4x + C$

Integration is the exact **opposite** of differentiation.

9.2 Rules For Integration

Therefore the rules for integrating are:
- add 1 on to the power
- divide by the new power
- add on the constant of integration

For example, integrate the following:

1. $\int (7x - 2)\, dx = \dfrac{7}{2}x^2 - 2x + C$

2. $\int (x^3 - 5x^2)\, dx = \dfrac{1}{4}x^4 - \dfrac{5}{3}x^3 + C$

3. $\int (6x - 6x^2 - \dfrac{2}{3}x^3)\, dx = 3x^2 - 2x^3 - \dfrac{1}{6}x^4 + C$

Exercise 9A: *Integrate the following:*

1. $\int (6x + 3)\, dx$

2. $\int (12x^2 + 4x - 2)\, dx$

3. $\int (16x^3 - 9x^2 - 6x + 3)\, dx$

4. $\int (7 - 2x)\, dx$

5. $\int (4x + 3x^2)\, dx$

6. $\int (6x^3 - 4x - 1)\, dx$

7. $\int (3x^4 - 2x^3 + 7x^2)\, dx$

8. $\int (\dfrac{2}{5}x^4 + 6x)\, dx$

9. $\int (\dfrac{3}{7}x^2 - 2x)\, dx$

10. $\int (\dfrac{5}{8}x^3 + \dfrac{1}{4}x^2 + x)\, dx$

11. $\int (6x - 4x^2 + 7)\, dx$

12. $\int (9 - 2x - 6x^2)\, dx$

13. $\int (10x + x^2 - 3)\, dx$

14. $\int (\dfrac{2}{3}x^3 + \dfrac{6}{5}x^2 - 1)\, dx$

Exercise 9A...

15. $\int (4x - \dfrac{2}{5}x^4 + 5)\, dx$

16. $\int (4x - 7)\, dx$

17. $\int (4x^3 - 7x^2 + 5x - 2)\, dx$

18. $\int (6x - \dfrac{2}{3}x^5)\, dx$

9.3 Integrating Expressions Where The x Term is on The Denominator

You must bring the x term to the numerator before integrating, using the same index rules you used for differentiating. For example, integrate the following:

1. $\int \dfrac{6}{x^2}\, dx$

Reorganise: $\dfrac{6}{x^2} = 6x^{-2}$

So: $\int 6x^{-2} = -6x^{-1} + C$

$\qquad\qquad = \dfrac{-6}{x} + C$

2. $\int (\dfrac{2}{5x^3} + 5x^3)\, dx$

Reorganise: $\dfrac{2}{5x^3} = \dfrac{2}{5}x^{-3}$

So: $\int (\dfrac{2}{5}x^{-3} + 5x^3)\, dx = \dfrac{-1}{5}x^{-2} + \dfrac{5}{4}x^4 + C$

$\qquad\qquad = \dfrac{-1}{5x^2} + \dfrac{5x^4}{4} + C$

Exercise 9B: *Integrate the following:*

1. $\int \dfrac{4}{x^2}\, dx$

2. $\int \dfrac{8}{x^3}\, dx$

3. $\int \dfrac{2}{x^5}\, dx$

4. $\int \dfrac{3}{5x^2}\, dx$

5. $\int \dfrac{2}{7x^3}\, dx$

6. $\int \dfrac{3}{5x^4}\, dx$

7. $\int (4x - \dfrac{4}{x^2})\, dx$

8. $\int (x^2 + 3 - \dfrac{2}{x^3})\, dx$

9. $\int (\dfrac{3x^2}{2} - \dfrac{2}{3x^2})\, dx$

10. $\int (6 + \dfrac{2}{4x^2})\, dx$

11. $\int (4x - \dfrac{1}{2x^5})\, dx$

12. $\int (\dfrac{5}{3x^2} + x^2)\, dx$

13. $\int (2x^3 - \dfrac{1}{2x^3})\, dx$

14. $\int (6x + 2 - \dfrac{4}{x^2})\, dx$

15. $\int (3x^2 - 5 + \dfrac{2}{5x^2})\, dx$

9.4 Definite Integration

Integrals such as all you have worked out so far are called **indefinite** integrals because of the unknown constant of integration, C. **Definite** integration is when you are given two numbers to substitute into your answer. You then subtract the totals, and in so doing the constant C cancels out leaving a definite integration. The rules for definite integration are as follows:

$$\int_a^b f(x)\, dx \qquad \text{where f(x) is a function of } x$$

• Integrate f(x),
• Substitute $x = b$ into the answer,
• Substitute $x = a$ into the answer,
• Subtract the two results.

For example, work out

1. $\int_1^4 (12x^2 + 4x - 2)\, dx$

$= [4x^3 + 2x^2 - 2x + C]_1^4$

$= [4\times4^3 + 2\times4^2 - 2\times4 + C] - [4\times1^3 + 2\times1^2 - 2\times1 + C]$

$= [256 + 32 - 8 + C] - [4 + 2 - 2 + C]$

$= [280] - [4]$

$= 276$

Because the 'C's will **always** cancel out, you could omit the 'C's when you integrate as in the next example.

2. $\int_{-1}^2 (\frac{3}{x^2} - 7x)\, dx$

Rearrange: $\int_{-1}^2 (3x^{-2} - 7x)\, dx$

$= [-3x^{-1} - \frac{7x^2}{2}]_{-1}^2$

$= [\frac{-3}{x} - \frac{7x^2}{2}]_{-1}^2$

$= [-15\frac{1}{2}] - [-\frac{1}{2}]$

$= -15$

Exercise 9C: *Work out the following:*

1. $\int_1^4 \frac{4}{x^2}\, dx$

2. $\int_{-1}^2 \frac{8}{x^3}\, dx$

3. $\int_1^2 \frac{2}{x^5}\, dx$

4. $\int_1^3 \frac{3}{5x^2}\, dx$

5. $\int_{-2}^1 \frac{2}{7x^3}\, dx$

6. $\int_1^2 \frac{3}{5x^4}\, dx$

7. $\int_{-1}^3 (4x - \frac{4}{x^2})\, dx$

8. $\int_2^4 (x^2 + 3 - \frac{2}{x^3})\, dx$

Exercise 9C...

9. $\int_{-2}^{-1} (\frac{3x^2}{2} - \frac{2}{3x^2})\, dx$

10. $\int_1^3 (6 + \frac{2}{4x^2})\, dx$

11. $\int_{-2}^1 (4x - \frac{1}{2x^5})\, dx$

12. $\int_{-2}^3 (\frac{5}{3x^2} + x^2)\, dx$

13. $\int_{-1}^1 (2x^3 - \frac{1}{2x^3})\, dx$

14. $\int_{-4}^{-2} (6x + 2 - \frac{4}{x^2})\, dx$

15. $\int_{-3}^2 (3x^2 - 5 + \frac{2}{5x^2})\, dx$

9.5 Finding y when given $\frac{dy}{dx}$

The rules are as follows:

• Integrate $\frac{dy}{dx}$,
• Substitute values of x and y to find C.
For example:

1. $\frac{dy}{dx} = 4x - 3$ Find:

(i) an expression for y in terms of x given that $x = 3$ when $y = -2$

Hence find (ii) y when $x = 6$

(i) Integrate: $y = \int(4x - 3)\, dx$
$= 2x^2 - 3x + C$

Substitute $x = 3$ and $y = -2$:
$y = 2x^2 - 3x + C$
$-2 = 2(3)^2 - 3(3) + C$
$-2 = 18 - 9 + C$
$-2 = 9 + C$
$-2 - 9 = C$
$-11 = C$

Thus $y = 2x^2 - 3x - 11$

(ii) Substitute $x = 6$ to get:
$y = 2(6)^2 - 3(6) - 11$
$y = 72 - 18 - 11$
$y = 43$

2. The gradient function of a curve is $6x^2 - 4x$ Find:
(i) the equation of the curve given that it passes through $(2, -3)$
(ii) hence find where the curve crosses the line $x = -1$

(i) The gradient function is $\frac{dy}{dx}$.

To find the equation of the curve you need to integrate $\frac{dy}{dx}$.

Integrate:
$$y = \int(6x^2 - 4x)\,dx$$
$$= 2x^3 - 2x^2 + C$$

Substitute $x = 2$ and $y = -3$ to give:

$$
\begin{aligned}
y &= 2x^3 - 2x^2 + C \\
-3 &= 2(2)^3 - 2(2)^2 + C \\
-3 &= 16 - 8 + C \\
-3 &= 8 + C \\
-3 - 8 &= C \\
-11 &= C
\end{aligned}
$$

Thus $\qquad y = 2x^3 - 2x^2 - 11$

(ii) Substitute $x = -1$ to get:

$$
\begin{aligned}
y &= 2(-1)^3 - 2(-1)^2 - 11 \\
y &= -2 - 2 - 11 \\
y &= -15
\end{aligned}
$$

Answer: $(-1, -15)$

Exercise 9D:

1. $\frac{dy}{dx} = 6x - 3$ Find:

(i) an expression for y in terms of x given that $x = 1$ when $y = 3$

(ii) hence find y when $x = 4$

2. $\frac{dy}{dx} = 4 + 2x$ Find:

(i) an expression for y in terms of x given that $x = -2$ when $y = 1$

(ii) hence find y when $x = 2$

3. $\frac{dy}{dx} = 3x^2 - 4x$ Find:

(i) an expression for y in terms of x given that $x = 1$ when $y = -4$

(ii) hence find y when $x = -3$

Exercise 9D...

4. $\frac{dy}{dx} = 6x + x^2$ Find:

(i) an expression for y in terms of x given that $x = -1$ when $y = -3$

(ii) hence find y when $x = 3$

5. The gradient function of a curve is $\frac{4}{x^2}$ Find:

(i) the equation of the curve given that it passes through $(1, 2)$

(ii) hence find where the curve crosses the line $x = -2$

6. The gradient function of a curve is $6x^2 - 2x + 5$ Find:

(i) the equation of the curve given that it passes through $(3, 1)$

(ii) hence find where the curve crosses the line $x = 4$

7. The gradient function of a curve is $3x + 4$ Find:

(i) the equation of the curve given that it passes through $(1, 2)$

(ii) hence find where the curve crosses the line $x = -2$

8. The gradient function of a curve is $6 - 2x + 3x^2$ Find:

(i) the equation of the curve given that it passes through $(-2, 4)$

(ii) hence find where the curve crosses the line $x = 6$

CHAPTER 10: AREA

10.1 The Area Under a Curve

Look at the curve below.

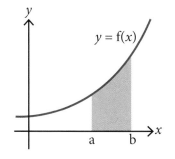

You can find the shaded area under the curve $y = f(x)$ between the lines $x = a$ and $x = b$ by integrating $y = f(x)$ between a and b:

$$\text{Area} = \int_a^b f(x)\,dx$$

The rules are as follows:

- Integrate the equation of the curve (ignoring C as the C's will cancel),
- Substitute the values $x = b$ and $x = a$ and subtract,
- Always give the positive value of the answer.

45

For example:

Find the area under the curve $y = x^2 - x - 12$ between the lines $x = 0$ and $x = 2$

$$\text{Area} \quad = \int_0^2 (x^2 - x - 12)\, dx$$

$$= {}_0^2\left[\frac{1}{3}x^3 - \frac{1}{2}x^2 - 12x\right]$$

$$= \left[\frac{1}{3}(8) - \frac{1}{2}(4) - 12(2)\right] - [0 - 0 - 0]$$

$$= \left[2\frac{2}{3} - 2 - 24\right] - [0]$$

$$= -23\frac{1}{3} - 0$$

$$= -23\frac{1}{3}$$

Answer: The area is $23\frac{1}{3}$

Note that:

• There are no units for the area.
• The curve $y = x^2 - x - 12$ is a quadratic curve and since the x^2 term is positive it will be a U shape. Thus the negative sign in the area confirms that the part of the curve between $x = 0$ and $x = 2$ is below the x axis.

Exercise 10A:

1. Find the area under the curve $y = x^2 - 7x + 10$ between the lines $x = 0$ and $x = 2$

2. Find the area under the curve $y = 12 - 7x + x^2$ between the lines $x = 0$ and $x = 2$

3. Find the area under the curve $y = 2x^2 + 7x - 4$ between the lines $x = -2$ and $x = 0$

4. Find the area under the curve $y = 3x^2 - 10x - 8$ between the lines $x = 0$ and $x = 3$

5. Find the area under the curve $y = x^2 + 2x - 24$ between the lines $x = 1$ and $x = 2$

6. Find the area under the curve $y = 2x^2 - 11x - 6$ between the lines $x = 2$ and $x = 4$

7. Find the area under the curve $y = x^2 + x - 6$ between the lines $x = 0$ and $x = 1$

8. Find the area under the curve $y = 4x^2 - 11x + 6$ between the lines $x = 0$ and $x = 2$

9. Find the area under the curve $y = 12 - 20x + 3x^2$ between the lines $x = 1$ and $x = 2$

10. Find the area under the curve $y = 3x^2 - 5x - 3$ between the lines $x = 0$ and $x = 2$

11. Find the area under the curve $y = 2x^2 - 3x - 20$ between the lines $x = 0$ and $x = 3$

Exercise 10A...

12. Find the area under the curve $y = 2x^2 + 9x - 5$ between the lines $x = -2$ and $x = 0$

13. Find the area under the curve $y = x^2 + 2x - 8$ between the lines $x = 0$ and $x = 2$

14. Find the area under the curve $y = 2x^2 - 5x - 25$ between the lines $x = 1$ and $x = 2$

15. Find the area under the curve $y = 3x^2 - 2x - 8$ between the lines $x = 0$ and $x = 2$

10.2 The Area Between a Quadratic Curve And The x–Axis

The rules are as follows:

• Find where the curve crosses the x axis by putting $y = 0$,
• Integrate the equation of the curve between these two values.

For example:

(i) Sketch the curve $y = x^2 + 3x - 10$
(ii) Find the area between the curve $y = x^2 + 3x - 10$ and the x axis.

The curve crosses the x axis when $y = 0$

So $\quad x^2 + 3x - 10 = 0$

$\qquad (x + 5)(x - 2) = 0$

Thus $\qquad x = -5$ and $x = 2$

The curve crosses the y axis when $x = 0$

So $\qquad y = (0)^2 + 3(0) - 10$

$\qquad y = -10$

You find the turning point by differentiating:

$$y = x^2 + 3x - 10$$

$$\frac{dy}{dx} = 2x + 3$$

$$\frac{dy}{dx} = 2x + 3 = 0$$

$$2x = -3$$

$$x = -1\frac{1}{2}$$

Since $\quad \dfrac{d^2y}{dx^2} = 2$ we know it is a minimum

Substituting $\quad x = -1\frac{1}{2}$ into $y = x^2 + 3x - 10$

gives $\qquad y = -12\frac{1}{4}$

You can then sketch the curve as follows:

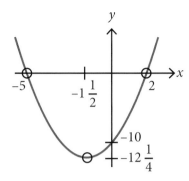

Area $= \int_{-5}^{2} (x^2 + 3x - 10)\, dx$

$= {}_{-5}^{2}[\frac{1}{3}x^3 + \frac{3}{2}x^2 - 10x]$

$= [\frac{1}{3}(8) + \frac{3}{2}(4) - 10(2)] - [\frac{1}{3}(-125) + \frac{3}{2}(25) - 10(-5)]$

$= [2\frac{2}{3} + 6 - 20] - [-41\frac{2}{3} + 37\frac{1}{2} + 50]$

$= [-11\frac{1}{3}] - [45\frac{5}{6}]$

$= -57\frac{1}{6}$

Answer: The area is $57\frac{1}{6}$

10.3 The Area Between a Cubic Curve And The x–Axis

The rules are similar for a cubic curve.
For example:
(i)   Sketch the curve $y = 2x^3 - 3x^2 - 72x$, giving answers to 2 decimal places where appropriate.

(ii) Find the area between the curve $y = 2x^3 - 3x^2 - 72x$ and the negative x axis.

The curve crosses the x axis when $y = 0$
So $y = 2x^3 - 3x^2 - 72x = 0$

$x(2x^2 - 3x - 72) = 0$

So either $x = 0$

or $2x^2 - 3x - 72 = 0$

You solve $2x^2 - 3x - 72 = 0$ by using the quadratic formula
$\dfrac{-b \pm \sqrt{b^2 - 4ac}}{2a}$:

Where $a = 2, b = -3, c = -72$

So $x = \dfrac{3 \pm \sqrt{(-3)^2 - (4 \times 2 \times -72)}}{2 \times 2}$

$x = \dfrac{3 \pm \sqrt{585}}{4}$

$x = \dfrac{3 + \sqrt{585}}{4}$ or $\dfrac{3 - \sqrt{585}}{4}$

So $x = 6.80$ or -5.30, to 2 decimal places.
Thus the curve crosses the x axis at $(0, 0)$, $(6.8, 0)$ and $(-5.30, 0)$

The turning points are when $\dfrac{dy}{dx} = 0$

$y = 2x^3 - 3x^2 - 72x$

$\dfrac{dy}{dx} = 6x^2 - 6x - 72 = 0$

Dividing through by 6 gives:

$x^2 - x - 12 = 0$
$(x - 4)(x + 3) = 0$
So either $x - 4 = 0$
$x = 4$
or $x + 3 = 0$
$x = -3$

You now find y by substituting $x = 4$ or -3 into the equation for y
$y = 2x^3 - 3x^2 - 72x$
When $x = 4$: $y = -208$
This turning point is at $(4, -208)$
When $x = -3$: $y = 135$
This turning point is at $(-3, 135)$

Now show what type each turning point is by getting $\dfrac{d^2y}{dx^2}$

$\dfrac{dy}{dx} = 6x^2 - 6x - 72$

$\dfrac{d^2y}{dx^2} = 12x - 6$

When $x = 4$: $\dfrac{d^2y}{dx^2} = 42$ and so it is a minimum

When $x = -3$: $\dfrac{d^2y}{dx^2} = -42$ and so it is a maximum

You can now sketch the curve:

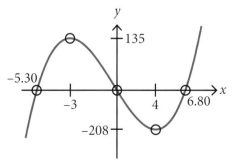

To find the area between the curve $y = 2x^3 - 3x^2 - 72x$ and the negative x-axis you must integrate $y = 2x^3 - 3x^2 - 72x$

between $x = -5.30$ and 0 as below:

$$\text{Area} = \int_{-5.30}^{0} (2x^3 - 3x^2 - 72x)\, dx$$

$$= {}_{-5.30}^{\quad 0}[\tfrac{1}{2}x^4 - x^3 - 36x^2]$$

$$= [0] - [394.524 + 148.877 - 1011.24]$$

$$= [0] - [-467.839]$$

$$= 467.84$$

Answer: The area is 467.84

Exercise 10B:

1. (i) Sketch the curve $y = x^2 - 8x + 12$
    (ii) Find the area between this curve and the x axis.
2. (i) Sketch the curve $y = x^2 - x - 20$
    (ii) Find the area between this curve and the x axis.
3. (i) Sketch the curve $y = -x^2 + x + 6$
    (ii) Find the area between this curve and the x axis.
4. (i) Sketch the curve $y = -2x^2 - 3x + 20$
    (ii) Find the area between this curve and the x axis.
5. (i) Sketch the curve $y = 2x^3 + 9x^2 - 60x$, giving answers to 2 decimal places where appropriate.
    (ii) Find the area between this curve and the positive x axis.

Exercise 10B...

6. (i) Sketch the curve $y = x^3 - 3x^2 - 24x$, giving answers to 2 decimal places where appropriate.
    (ii) Find the area between this curve and the negative x axis.
7. (i) Sketch the curve $y = x^3 - 3x^2 - 45x$, giving answers to 2 decimal places where appropriate.
    (ii) Find the area between this curve and the negative x axis.
8. (i) Sketch the curve $y = 2x^3 + 15x^2 - 36x$, giving answers to 2 decimal places where appropriate.
    (ii) Find the area between this curve and the positive x axis.
9. (i) Sketch the curve $y = x^3 - 48x$, giving answers to 2 decimal places where appropriate.
    (ii) Find the area between this curve and the positive x axis.
10. (i) Sketch the curve $y = 2x^3 + 9x^2 - 108x$, giving answers to 2 decimal places where appropriate.
    (ii) Find the area between this curve and the positive x axis.
11. (i) Sketch the curve $y = -x^3 + 6x^2 + 135x$.
    (ii) Find the area between this curve and the negative x axis.

CHAPTER 11: MATRICES

11.1 Matrices

A matrix is a rectangular collection of numbers. For example:

1. $\begin{pmatrix} 4 & -3 \\ -6 & 2 \end{pmatrix}$ is a matrix with 2 rows and 2 columns.

This is called a 2×2 matrix.

....

2. $\begin{pmatrix} 4 & -3 & 5 \end{pmatrix}$ is a matrix with 1 row and 3 columns.

This is called a 1×3 matrix.

....

3. $\begin{pmatrix} 4 \\ 2 \\ -5 \\ -4 \end{pmatrix}$ is a matrix with 4 rows and 1 column.

This is called a 4×1 matrix.

11.2 Adding And Subtracting Matrices

You can only add matrices which have the same number of rows as columns. The rule for adding or subtracting matrices is:

• Add or subtract numbers in corresponding places

For example:

1. $\begin{pmatrix} 6 & -2 \\ -3 & 4 \end{pmatrix} + \begin{pmatrix} -7 & 4 \\ -4 & 2 \end{pmatrix} = \begin{pmatrix} -1 & 2 \\ -7 & 6 \end{pmatrix}$

....

2. $\begin{pmatrix} -4 & -1 \\ 7 & 5 \end{pmatrix} - \begin{pmatrix} -3 & 8 \\ -1 & 9 \end{pmatrix} = \begin{pmatrix} -1 & -9 \\ 8 & -4 \end{pmatrix}$

....

3. $\begin{pmatrix} 2x & -y \\ 6 & -3v \end{pmatrix} + \begin{pmatrix} x & -5y \\ -3 & 2v \end{pmatrix} = \begin{pmatrix} 3x & -6y \\ 3 & -v \end{pmatrix}$

....

4. $A = \begin{pmatrix} 5 & 3 \\ -2 & -4 \end{pmatrix}$ and $B = \begin{pmatrix} -2 & 6 \\ -5 & -3 \end{pmatrix}$

$A+B = \begin{pmatrix} 5 & 3 \\ -2 & -4 \end{pmatrix} + \begin{pmatrix} -2 & 6 \\ -5 & -3 \end{pmatrix} = \begin{pmatrix} 3 & 9 \\ -7 & -7 \end{pmatrix}$

Exercise 11A:

1. $\begin{pmatrix} -3 \\ 4 \end{pmatrix} + \begin{pmatrix} 2 \\ -1 \end{pmatrix}$

2. $\begin{pmatrix} -6 \\ 1 \end{pmatrix} - \begin{pmatrix} -2 \\ 5 \end{pmatrix}$

3. $(-1 \;\; 4) + (5 \;\; -2)$

4. $(2 \;\; -3) - (-1 \;\; -2)$

5. $\begin{pmatrix} x \\ -3 \end{pmatrix} + \begin{pmatrix} -2x \\ 1 \end{pmatrix}$

6. $(4 \;\; -3y) - (2 \;\; y)$

7. $\begin{pmatrix} -2x \\ 3y \end{pmatrix} + \begin{pmatrix} x \\ -4y \end{pmatrix}$

8. $(-3x \;\; y) - (x \;\; -2y)$

9. $\begin{pmatrix} 1 & -5 \\ 9 & -4 \end{pmatrix} + \begin{pmatrix} -6 & -5 \\ 5 & 2 \end{pmatrix}$

10. $\begin{pmatrix} -4 & -8 \\ -3 & 4 \end{pmatrix} - \begin{pmatrix} -4 & 7 \\ -2 & -7 \end{pmatrix}$

11. $\begin{pmatrix} x & -2y \\ 8 & 4t \end{pmatrix} + \begin{pmatrix} 3x & 5y \\ -2 & -2t \end{pmatrix}$

12. $\begin{pmatrix} 6 & -2 \\ -3 & 4 \end{pmatrix} + \begin{pmatrix} -7 & 4 \\ -4 & 2 \end{pmatrix}$

13. $\begin{pmatrix} 3n & -2m \\ -3 & -6q \end{pmatrix} - \begin{pmatrix} -7n & m \\ 11 & -3q \end{pmatrix}$

14. $\begin{pmatrix} 4.6 & -2.6 \\ -3.4 & 4.7 \end{pmatrix} + \begin{pmatrix} -7.4 & 4.8 \\ -4.1 & 2.8 \end{pmatrix}$

15. $\begin{pmatrix} -2.6 & 3.5 \\ -3.1 & -4.6 \end{pmatrix} - \begin{pmatrix} -7.5 & -5.4 \\ -4.8 & 7.2 \end{pmatrix}$

16. $\begin{pmatrix} -3 & -8 \\ 9 & -4 \end{pmatrix} + \begin{pmatrix} -2 & -4 \\ -7 & 3 \end{pmatrix} - \begin{pmatrix} 3 & -7 \\ -7 & -4 \end{pmatrix}$

17. $A = \begin{pmatrix} -6 & 8 \\ -3 & 4 \end{pmatrix}$ $B = \begin{pmatrix} 4 & 7 \\ -2 & 6 \end{pmatrix}$

$C = \begin{pmatrix} -2 & -9 \\ 8 & 6 \end{pmatrix}$ $D = \begin{pmatrix} -4 & 8 \\ -6 & -2 \end{pmatrix}$

Find (i) A + B (ii) B – A (iii) A + C (iv) D – A (v) B + C (vi) D – C (vii) A + D (viii) C – B

11.3 Multipling And Dividing Matrices By Whole Numbers

You multiply or divide every number in the matrix by the whole number.

For example:

1. $P = \begin{pmatrix} 4 & -7 \\ -6 & 2 \end{pmatrix}$ Find (i) 6P (ii) $\frac{-1}{2}$ P

(i) Multiplying all the numbers by 6 gives:

$6P = \begin{pmatrix} 24 & -42 \\ -36 & 12 \end{pmatrix}$

(ii) Dividing all the numbers by –2 gives:

$\frac{-1}{2} P = \begin{pmatrix} -2 & 3.5 \\ 3 & -1 \end{pmatrix}$

....

2. $M = \begin{pmatrix} 5 & -8 \\ -4 & 5 \end{pmatrix}$ and $N = \begin{pmatrix} -2 & -6 \\ 6 & -2 \end{pmatrix}$

Find (i) 2M + 3N (ii) $4M - \frac{1}{2}N$

(i) $\begin{pmatrix} 10 & -16 \\ -8 & 10 \end{pmatrix} + \begin{pmatrix} -6 & -18 \\ 18 & -6 \end{pmatrix} = \begin{pmatrix} 4 & -34 \\ 10 & 4 \end{pmatrix}$

(ii) $\begin{pmatrix} 20 & -32 \\ -16 & 20 \end{pmatrix} - \begin{pmatrix} -1 & -3 \\ 3 & -1 \end{pmatrix} = \begin{pmatrix} 21 & -29 \\ -19 & 21 \end{pmatrix}$

Exercise 11B:

$A = \begin{pmatrix} 5 & 6 \\ -2 & 4 \end{pmatrix}$ $B = \begin{pmatrix} -3 & -2 \\ 6 & -5 \end{pmatrix}$ $C = \begin{pmatrix} 1 & -4 \\ 3 & -6 \end{pmatrix}$

$D = \begin{pmatrix} 3 & -5 \\ -5 & 3 \end{pmatrix}$ $E = \begin{pmatrix} -9 & 8 \\ 3 & -6 \end{pmatrix}$

Find:

1. 5A
2. 3B
3. –5C
4. 8D
5. 7E
6. $\frac{1}{2}$ A
7. $\frac{1}{10}$ D
8. $\frac{1}{2}$ B
9. 2A + B
10. 3C – D
11. 2E + 4C
12. 5B – 3A
13. 2A + D – E
14. 3E – 2C + A
15. A + B – C – D

16. $A = \begin{pmatrix} -2 \\ 5 \end{pmatrix}$ $B = (3 \;\; -4)$ $C = \begin{pmatrix} 2 \\ -7 \end{pmatrix}$ $D = (-1 \;\; -5)$

Find: (i) 3A (ii) 2B (iii) 4C (iv) –2D

(v) $\frac{1}{2}$A (vi) $\frac{3}{4}$B (vii) 3(A – C) (viii) $\frac{1}{2}$(B + D)

11.4 Solving Matrix Equations

You follow the same rules in solving matrix equations as in solving algebraic linear equations, ie you separate the variable from the numbers.

For example, solve the following:

1. $X + \begin{pmatrix} -2 & 7 \\ 5 & -3 \end{pmatrix} = \begin{pmatrix} 4 & -1 \\ 2 & -5 \end{pmatrix}$

Subtracting $\begin{pmatrix} -2 & 7 \\ 5 & -3 \end{pmatrix}$ from both sides gives :

$X = \begin{pmatrix} 4 & -1 \\ 2 & -5 \end{pmatrix} - \begin{pmatrix} -2 & 7 \\ 5 & -3 \end{pmatrix}$

Then subtract each corresponding entry as before to get:

$X = \begin{pmatrix} 6 & -8 \\ -3 & -2 \end{pmatrix}$

2. $\begin{pmatrix} 2 & -4 \\ 5 & -3 \end{pmatrix} - 2X = \begin{pmatrix} 9 & 4 \\ 6 & -5 \end{pmatrix}$

Adding 2X to both sides gives:

$\begin{pmatrix} 2 & -4 \\ 5 & -3 \end{pmatrix} = \begin{pmatrix} 9 & 4 \\ 6 & -5 \end{pmatrix} + 2X$

Subtracting $\begin{pmatrix} 9 & 4 \\ 6 & -5 \end{pmatrix}$ from both sides gives :

$\begin{pmatrix} 2 & -4 \\ 5 & -3 \end{pmatrix} - \begin{pmatrix} 9 & 4 \\ 6 & -5 \end{pmatrix} = 2X$

Then subtract each corresponding entry as before to get

$2X = \begin{pmatrix} -7 & -8 \\ -1 & 2 \end{pmatrix}$

You then divide every entry by 2 to get

$X = \begin{pmatrix} -3.5 & -4 \\ -0.5 & 1 \end{pmatrix}$

Exercise 11C: *Solve:*

1. $X - \begin{pmatrix} -2 & 7 \\ 5 & 3 \end{pmatrix} = \begin{pmatrix} 4 & -2 \\ 5 & 9 \end{pmatrix}$

2. $X + \begin{pmatrix} -3 & 5 \\ 4 & -2 \end{pmatrix} = \begin{pmatrix} 6 & -2 \\ 9 & -4 \end{pmatrix}$

3. $X + (-7 \ 4 \ 5) = (9 \ -2 \ 6)$

4. $X - \begin{pmatrix} -7 \\ 2 \end{pmatrix} = \begin{pmatrix} 4 \\ -6 \end{pmatrix}$

5. $2X - \begin{pmatrix} 3 & 4 \\ -5 & 2 \end{pmatrix} = \begin{pmatrix} 7 & -2 \\ 9 & -6 \end{pmatrix}$

6. $3X + \begin{pmatrix} -2 & 5 \\ 6 & -1 \end{pmatrix} = \begin{pmatrix} 7 & -1 \\ 9 & -7 \end{pmatrix}$

7. $\begin{pmatrix} 4 & -2 \\ 6 & 3 \end{pmatrix} + 4X = \begin{pmatrix} 8 & -6 \\ 2 & 15 \end{pmatrix}$

Exercise 11C...

8. $\begin{pmatrix} 7 & -3 \\ -5 & 4 \end{pmatrix} - 2X = \begin{pmatrix} 3 & -9 \\ 7 & -10 \end{pmatrix}$

9. $5X + \begin{pmatrix} -2 & 7 \\ 5 & -3 \end{pmatrix} = \begin{pmatrix} 8 & 2 \\ -10 & -8 \end{pmatrix}$

10. $\begin{pmatrix} 7 & 4 \\ -3 & 2 \end{pmatrix} - 2X = \begin{pmatrix} 1 & -2 \\ 5 & -4 \end{pmatrix}$

11.5 Multiplying Matrices By Matrices

Two matrices A and B can only be multiplied together to give AB if the number of columns in A equals the number of rows in B. The **order** of the answer (ie the number of rows and columns in the matrix) will be the number of rows in A × the number of columns in B. The rule for multiplying two matrices is to multiply each row into each column by:

• multiplying each pair of numbers in turn and then
• adding these answers

For example:

1. $\begin{pmatrix} -2 & 3 \\ 4 & 2 \end{pmatrix} \begin{pmatrix} 1 \\ 4 \end{pmatrix} = \begin{pmatrix} -2 \times 1 + 3 \times 4 \\ 4 \times 1 + 2 \times 4 \end{pmatrix} = \begin{pmatrix} -2 + 12 \\ 4 + 8 \end{pmatrix} = \begin{pmatrix} 10 \\ 12 \end{pmatrix}$

$\quad$ 2×2 $\qquad$ 2×1 $\ =\ $ $\qquad$ 2×1

2. $(-3 \ 4) \begin{pmatrix} 2 \\ -5 \end{pmatrix} = (-3 \times 2 + 4 \times -5) = (-6 - 20) = (-26)$

$\quad$ 1×2 $\quad$ 2×1 $\ =\ $ $\qquad$ 1×1

Note: You must put brackets round the −26 as it is a matrix.

3. $\begin{pmatrix} 1 \\ 4 \end{pmatrix} (-2 \ -3) = \begin{pmatrix} 1 \times -2 & 1 \times -3 \\ 4 \times -2 & 4 \times -3 \end{pmatrix} = \begin{pmatrix} -2 & -3 \\ -8 & -12 \end{pmatrix}$

$\quad$ 2×1 $\quad$ 1×2 $\ =\ $ $\qquad$ 2×2

4. $A = \begin{pmatrix} -3 & 4 \\ 5 & 2 \end{pmatrix}$ $B = \begin{pmatrix} 1 & -2 \\ 3 & -4 \end{pmatrix}$

Find: (i) AB (ii) BA

(i) $\begin{pmatrix} -3 & 4 \\ 5 & 2 \end{pmatrix} \begin{pmatrix} 1 & -2 \\ 3 & -4 \end{pmatrix} = \begin{pmatrix} -3+12 & 6-16 \\ 5+6 & -10-8 \end{pmatrix} = \begin{pmatrix} 9 & -10 \\ 11 & -18 \end{pmatrix}$

$\quad$ 2×2 $\qquad$ 2×2 $\ =\ $ $\qquad$ 2×2

(ii) $\begin{pmatrix} 1 & -2 \\ 3 & -4 \end{pmatrix} \begin{pmatrix} -3 & 4 \\ 5 & 2 \end{pmatrix} = \begin{pmatrix} -3-10 & 4-4 \\ -9-20 & 12-8 \end{pmatrix} = \begin{pmatrix} -13 & 0 \\ -29 & 4 \end{pmatrix}$

$\quad$ 2×2 $\qquad$ 2×2 $\ =\ $ $\qquad$ 2×2

Note that AB ≠ BA

5. $C = \begin{pmatrix} -2 & 3 \\ 5 & 2 \end{pmatrix}$ Find C^2

$$C^2 = \begin{pmatrix} -2 & 3 \\ 5 & 2 \end{pmatrix} \begin{pmatrix} -2 & 3 \\ 5 & 2 \end{pmatrix} = \begin{pmatrix} 4+15 & -6+6 \\ -10+10 & 15+4 \end{pmatrix} = \begin{pmatrix} 19 & 0 \\ 0 & 19 \end{pmatrix}$$

Note that there is **no** shortcut in finding C^2. You do **not** square each entry.

Exercise 11D:

1. $(-2 \ 4) \begin{pmatrix} 5 \\ -3 \end{pmatrix}$

6. $\begin{pmatrix} -1 & 4 \\ 5 & 2 \end{pmatrix} \begin{pmatrix} 2 \\ -3 \end{pmatrix}$

2. $\begin{pmatrix} 4 \\ 2 \end{pmatrix} (-2 \ 4)$

7. $(-2 \ -5) \begin{pmatrix} -2 \\ 4 \end{pmatrix}$

3. $\begin{pmatrix} -2 & 4 \\ 5 & 2 \end{pmatrix} \begin{pmatrix} 1 \\ 4 \end{pmatrix}$

8. $\begin{pmatrix} -3 \\ 5 \end{pmatrix} (4 \ -2)$

4. $(-1 \ -3) \begin{pmatrix} -5 \\ 2 \end{pmatrix}$

9. $\begin{pmatrix} 3 & -2 \\ -1 & 4 \end{pmatrix} \begin{pmatrix} -2 \\ 7 \end{pmatrix}$

5. $\begin{pmatrix} 2 \\ -3 \end{pmatrix} (-5 \ 3)$

10. $A = \begin{pmatrix} -2 & 5 \\ 4 & 3 \end{pmatrix}$ $B = \begin{pmatrix} 1 & -2 \\ -3 & 5 \end{pmatrix}$ $C = \begin{pmatrix} -2 & 5 \\ 3 & -4 \end{pmatrix}$

Find (i) AB (ii) AC (iii) BC (iv) BA
(v) CA (vi) CB (vii) A^2 (viii) B^2 (ix) C^2

11.6 Determinant Of A Matrix

The determinant of a matrix is a number calculated as follows:

If $A = \begin{pmatrix} a & b \\ c & d \end{pmatrix}$ then the determinant of A is given by:

det A = ad – bc

You must **not** put brackets around this answer as it is a number and not a matrix. For example:

Calculate the determinant of A, if $A = \begin{pmatrix} 3 & -8 \\ 7 & -4 \end{pmatrix}$

$$\begin{aligned} \text{det A} &= 3 \times (-4) - (-8) \times 7 \\ &= -12 - (-56) \\ &= -12 + 56 \\ &= 44 \end{aligned}$$

Exercise 11E: *Find the determinant of the following matrices:*

1. $\begin{pmatrix} 5 & 7 \\ 4 & 2 \end{pmatrix}$

6. $\begin{pmatrix} 4 & 5 \\ -1 & -3 \end{pmatrix}$

2. $\begin{pmatrix} 8 & -6 \\ 3 & 5 \end{pmatrix}$

7. $\begin{pmatrix} 5 & -4 \\ -2 & 5 \end{pmatrix}$

3. $\begin{pmatrix} -4 & 5 \\ 3 & -3 \end{pmatrix}$

8. $\begin{pmatrix} 4 & -3 \\ -2 & -4 \end{pmatrix}$

4. $\begin{pmatrix} 7 & -6 \\ -5 & -2 \end{pmatrix}$

9. $\begin{pmatrix} 5 & -4 \\ -3 & 6 \end{pmatrix}$

5. $\begin{pmatrix} 6 & -6 \\ 3 & 9 \end{pmatrix}$

10. $\begin{pmatrix} 4 & -4 \\ -3 & -1 \end{pmatrix}$

11.7 The Unit Matrix

The unit matrix is called I (for **identity** matrix). It is defined as:

$$I = \begin{pmatrix} 1 & 0 \\ 0 & 1 \end{pmatrix}$$

When any matrix A is multiplied by I, or when I is multiplied by any matrix A, the answer is always A. In other words, AI = IA = A . For example:

If $A = \begin{pmatrix} 3 & -5 \\ -2 & 6 \end{pmatrix}$ then:

$$AI = \begin{pmatrix} 3 & -5 \\ -2 & 6 \end{pmatrix} \begin{pmatrix} 1 & 0 \\ 0 & 1 \end{pmatrix} = \begin{pmatrix} 3 & -5 \\ -2 & 6 \end{pmatrix}$$

$$IA = \begin{pmatrix} 1 & 0 \\ 0 & 1 \end{pmatrix} \begin{pmatrix} 3 & -5 \\ -2 & 6 \end{pmatrix} = \begin{pmatrix} 3 & -5 \\ -2 & 6 \end{pmatrix}$$

11.8 Inverse Of A Matrix

The inverse of any matrix A is called A^{-1}. Whenever any matrix A is multiplied by its inverse A^{-1}, or vice–versa, the answer is always I. In other words, $AA^{-1} = A^{-1}A = I$. The inverse of a matrix is calculated as follows:

If $A = \begin{pmatrix} a & b \\ c & d \end{pmatrix}$ then the inverse of A is given by:

$$A^{-1} = \frac{1}{\text{det A}} \begin{pmatrix} d & -b \\ -c & a \end{pmatrix}$$

For example:
1. Find the inverse of the following matrices

(i) $A = \begin{pmatrix} 5 & 7 \\ -4 & 3 \end{pmatrix}$

det A = 15 – (–28) = 15+28 = 43

$$A^{-1} = \frac{1}{43} \begin{pmatrix} 3 & -7 \\ 4 & 5 \end{pmatrix}$$

You should **not** work out the bracket.

(ii) B = $\begin{pmatrix} -3 & -2 \\ -5 & -3 \end{pmatrix}$

det B = 9 − 10 = −1

$B^{-1} = \dfrac{1}{1} \begin{pmatrix} -3 & 2 \\ 5 & -3 \end{pmatrix} = -1 \begin{pmatrix} -3 & 2 \\ 5 & -3 \end{pmatrix}$

$= \begin{pmatrix} 3 & -2 \\ -5 & 3 \end{pmatrix}$

You **should** work out the bracket when the term outside is −1.

....

2. Show that the matrix C = $\begin{pmatrix} 5 & -4 \\ 10 & -8 \end{pmatrix}$ has no inverse.

det C = −40 − (−40) = 0

$C^{-1} = \dfrac{1}{0} \begin{pmatrix} -8 & 4 \\ -10 & 5 \end{pmatrix}$

You cannot divide 1 by 0. Therefore C has no inverse.

Exercise 11F: *Find the inverse of the following matrices*

1. $\begin{pmatrix} 3 & -6 \\ 7 & 2 \end{pmatrix}$ 2. $\begin{pmatrix} 3 & -2 \\ 5 & -4 \end{pmatrix}$

3. $\begin{pmatrix} -2 & 4 \\ 5 & 2 \end{pmatrix}$ 4. $\begin{pmatrix} -3 & 5 \\ 4 & -2 \end{pmatrix}$

5. $\begin{pmatrix} -2 & 5 \\ 3 & -4 \end{pmatrix}$ 6. $\begin{pmatrix} 5 & -2 \\ -8 & 3 \end{pmatrix}$

7. $\begin{pmatrix} -2 & 6 \\ 4 & 3 \end{pmatrix}$ 8. $\begin{pmatrix} 1 & -3 \\ 5 & 4 \end{pmatrix}$

9. $\begin{pmatrix} -3 & 4 \\ 5 & 2 \end{pmatrix}$ 10. $\begin{pmatrix} -2 & -3 \\ 3 & 5 \end{pmatrix}$

11. $\begin{pmatrix} -5 & 4 \\ 3 & -2 \end{pmatrix}$ 12. $\begin{pmatrix} 1 & -2 \\ 5 & 3 \end{pmatrix}$

13. $\begin{pmatrix} -2 & 4 \\ -5 & 3 \end{pmatrix}$ 14. $\begin{pmatrix} -1 & 5 \\ 7 & -2 \end{pmatrix}$

15. $\begin{pmatrix} 6 & -2 \\ 7 & 4 \end{pmatrix}$ 16. $\begin{pmatrix} -2 & 5 \\ 4 & -3 \end{pmatrix}$

17. Explain why the matrix $\begin{pmatrix} -2 & -5 \\ 4 & 10 \end{pmatrix}$ has no inverse.

18. Explain why the matrix $\begin{pmatrix} 8 & -6 \\ -4 & 3 \end{pmatrix}$ has no inverse.

19. The matrix $\begin{pmatrix} 4 & a \\ 6 & 9 \end{pmatrix}$ has no inverse. Find a.

20. The matrix $\begin{pmatrix} -3 & 5 \\ 4 & q \end{pmatrix}$ has no inverse. Find q.

11.9 Solving Harder Matrix Equations

You can use the inverse of a matrix to solve matrix equations of the form AX = B as follows:

• Find the inverse of A
• Multiply A^{-1} by B **in this order**

This method works because:

AX = B

Multiplying both sides by A^{-1} gives:

A^{-1} AX = A^{-1} B

But A^{-1} A = I

So   IX = A^{-1} B

But   IX = X

So   X = A^{-1} B

For example, if:

A = $\begin{pmatrix} -2 & 5 \\ 3 & -4 \end{pmatrix}$ B = $\begin{pmatrix} 1 & -2 \\ 4 & -6 \end{pmatrix}$ C = $\begin{pmatrix} -3 \\ 7 \end{pmatrix}$

Solve (i) AX = C (ii) BX = A

(i)   AX = C

        X = A^{-1} C

        det A = 8 − 15 = −7

So $A^{-1} = \dfrac{-1}{7} \begin{pmatrix} -4 & -5 \\ -3 & -2 \end{pmatrix}$

So $X = \dfrac{-1}{7} \begin{pmatrix} -4 & -5 \\ -3 & -2 \end{pmatrix} \begin{pmatrix} -3 \\ 7 \end{pmatrix}$

        $\quad\quad\quad 2\times2 \quad\quad 2\times1$

        $= \dfrac{-1}{7} \begin{pmatrix} 12 - 35 \\ 9 - 14 \end{pmatrix}$

        $\quad\quad\quad\quad 2\times1$

        $= \dfrac{-1}{7} \begin{pmatrix} -23 \\ -5 \end{pmatrix}$

You can leave your answer in this form.

(ii)   BX = A

        X = B^{-1} A

        det B = −6 + 8 = 2

So $B^{-1} = \dfrac{1}{2} \begin{pmatrix} -6 & 2 \\ -4 & 1 \end{pmatrix}$

So $X = \dfrac{1}{2} \begin{pmatrix} -6 & 2 \\ -4 & 1 \end{pmatrix} \begin{pmatrix} -2 & 5 \\ 3 & -4 \end{pmatrix}$

        $\quad\quad\quad 2\times2 \quad\quad 2\times2$

        $= \dfrac{1}{2} \begin{pmatrix} 18 & -38 \\ 11 & -24 \end{pmatrix}$

        $\quad\quad\quad\quad 2\times2$

        $= \begin{pmatrix} 9 & -19 \\ 5\tfrac{1}{2} & -12 \end{pmatrix}$

Exercise 11G:

$$A = \begin{pmatrix} 2 & 5 \\ 4 & 3 \end{pmatrix} \quad B = \begin{pmatrix} 4 & -2 \\ 5 & -4 \end{pmatrix} \quad C = \begin{pmatrix} -2 & 7 \\ 3 & -4 \end{pmatrix}$$

$$D = \begin{pmatrix} -7 \\ 4 \end{pmatrix} \quad E = \begin{pmatrix} 1 \\ -3 \end{pmatrix}$$

Solve the following:

1. AX = D
2. BX = E
3. CX = D
4. AX = E
5. AX = B
6. CX = A
7. BX = C
8. AX = C
9. CX = B
10. BX = A
11. AX + B = C
12. BX – D = E
13. CX –2A = 3B
14. AX + 3D = 2E

11.10 Solving Two Simultaneous Equations Using Matrices

You can use matrices to solve simultaneous equations as follows:

- Rewrite the two equations as one matrix equation, ie AX = B
- A is the matrix with the coefficients of x and y from each equation,
- $X = \begin{pmatrix} x \\ y \end{pmatrix}$
- B is the matrix with the constants from each equation,
- Find the inverse of A
- Multiply A^{-1} by B **in this order**,
- Write out the answers for x and y.

For example:

Solve the following simultaneous equations using matrices:

(i) $4x - 3y = 19$

$2x + 5y = -10$

$$A = \begin{pmatrix} 4 & -3 \\ 2 & 5 \end{pmatrix} \quad X = \begin{pmatrix} x \\ y \end{pmatrix} \quad B = \begin{pmatrix} 19 \\ -10 \end{pmatrix}$$

$$AX = B$$
$$X = A^{-1} B$$
$$\det A = 20 - (-6) = 20 + 6 = 26$$

So $$A^{-1} = \frac{1}{26} \begin{pmatrix} 5 & 3 \\ -2 & 4 \end{pmatrix}$$

So
$$X = \frac{1}{26} \begin{pmatrix} 5 & 3 \\ -2 & 4 \end{pmatrix} \begin{pmatrix} 19 \\ -10 \end{pmatrix}$$
$$= \frac{1}{26} \begin{pmatrix} 95 - 30 \\ -38 - 40 \end{pmatrix}$$
$$= \frac{1}{26} \begin{pmatrix} 65 \\ -78 \end{pmatrix}$$
$$= \begin{pmatrix} 2\frac{1}{2} \\ -3 \end{pmatrix}$$

Therefore $x = 2\frac{1}{2}$ and $y = -3$

(ii) $\quad 3x + y - 6 = 0$

$\quad x - 2y - 9 = 0$

You must first re-write the simultaneous equations:

$$3x + y = 6$$
$$x - 2y = 9$$

$$A = \begin{pmatrix} 3 & 1 \\ 1 & -2 \end{pmatrix} \quad X = \begin{pmatrix} x \\ y \end{pmatrix} \quad B = \begin{pmatrix} 6 \\ 9 \end{pmatrix}$$

$$AX = B$$
$$X = A^{-1} B$$
$$\det A = -6 - 1 = -7$$

So $$A^{-1} = \frac{-1}{7} \begin{pmatrix} -2 & -1 \\ -1 & 3 \end{pmatrix}$$

So
$$X = \frac{-1}{7} \begin{pmatrix} -2 & -1 \\ -1 & 3 \end{pmatrix} \begin{pmatrix} 6 \\ 9 \end{pmatrix}$$
$$= \frac{-1}{7} \begin{pmatrix} -12 - 9 \\ -6 + 27 \end{pmatrix}$$
$$= \frac{-1}{7} \begin{pmatrix} -21 \\ 21 \end{pmatrix}$$
$$= \begin{pmatrix} 3 \\ -3 \end{pmatrix}$$

Therefore $x = 3$ and $y = -3$

Exercise 11H: *Solve the following simultaneous equations using matrices:*

1. $2x + 3y = 2$ $\qquad$ $x + 4y = -4$
2. $4x - 3y = -23$ $\qquad$ $x + 2y = 8$
3. $2x + 5y = -4$ $\qquad$ $-3x - y = -7$
4. $4x - 2y = -2$ $\qquad$ $2x + 7y = 31$
5. $3x - 5y - 11 = 0$ $\qquad$ $x + 4y + 19 = 0$
6. $2x - 9y = 40$ $\qquad$ $3x + y = 2$
7. $4x - 3y = -24$ $\qquad$ $x - 5y = -23$
8. $6x + 3y = -3$ $\qquad$ $2x - 5y = -37$
9. $2x + 3y - 4 = 0$ $\qquad$ $x - 5y - 15 = 0$
10. $4x - 2y = 6$ $\qquad$ $3x + 9y = -69$
11. $3x - y = 1$ $\qquad$ $2x + 3y = 19$
12. $2x + 4y = 24$ $\qquad$ $x - 5y = -9$
13. $-2x + 5y = -18$ $\qquad$ $3x - 4y = 20$
14. $-x + 2y = 13$ $\qquad$ $-2x - 3y = -9$

Exercise 11H...

15. $4x - y = -5$	$2x + 3y = -13$	**21.** $3x + 2y = -6$	$2x - 5y = 34$
16. $2x - 5y = 18$	$-x + 4y = -12$	**22.** $4x + 3y = -6$	$x - 2y = -7$
17. $3x - 2y = -9$	$2x + 5y = 13$	**23.** $2x - 5y = 22$	$3x - 2y = 11$
18. $3x + 4y = -22$	$-2x - 5y = 24$	**24.** $-4x + 3y = -1$	$-2x - 5y = 19$
19. $2x - 7y = 20$	$3x + 8y = -7$	**25.** $2x - 7y = 24$	$3x + 2y = 11$
20. $-4x + 2y = 4$	$2x - 5y = 14$		

CHAPTER 12: LOGARITHMS

12.1 Definition

Consider the following: $10^2 = 100$

10 is called the **base**

2 is called the **index**

The logarithm of a number is defined as: **The power to which you would raise the base to get the number.**
So you could define $\log_{10} 100 = 2$ since $10^2 = 100$

For example:

$$\log_2 8 = 3$$
since
$$2^3 = 8$$
$$\log_4 0.25 = -1$$
since
$$4^{-1} = \frac{1}{4} = 0.25$$
$$\log_7 7 = 1$$
since
$$7^1 = 7$$

Note that $\log_a a = 1$ for any number, a, since $a^1 = 1$

Exercise 12A: *Work out:*

1. $\log_2 4$	**6.** $\log_3 1$
2. $\log_2 32$	**7.** $\log_4 16$
3. $\log_2 0.5$	**8.** $\log_5 125$
4. $\log_3 27$	**9.** $\log_{10} 1000000$
5. $\log_3 81$	**10.** $\log_{10} 0.001$

12.2 Logs To Base 10

Logs to base 10 are generally written without the base highlighted, ie log 4 means $\log_{10} 4$. All other bases must be specifically written.

Since logs are simply powers or indices to which you

would raise the base to get the number, then the laws of indices apply to the laws of logs, ie:

Law 1 $\log(ab) = \log a + \log b$

Law 2 $\log(\frac{a}{b}) = \log a - \log b$

Law 3 $\log a^n = n \log a$

For example:

Write the following in terms x, y and/or z where $\log a = x$, $\log b = y$ and $\log c = z$

1. $\log bc$ 2. $\log \frac{c}{b}$ 3. $\log a^7$ 4. $\log \frac{a^4 b^2}{c^3}$ 5. $\log \sqrt{\frac{a}{c}}$

You can simplify these logarithms by rewriting them using the 3 laws above.

1. $\log bc = \log b + \log c$ (using law 1)
So $\log bc = y + z$

2. $\log \frac{c}{b} = \log c - \log b$ (using law 2)
So $\log \frac{c}{b} = z - y$

3. $\log a^7 = 7 \log a$ (using law 3)
So $\log a^7 = 7x$

4. $\log \frac{a^4 b^2}{c^3} = \log a^4 + \log b^2 - \log c^3$
(using laws 1 and 2)
$= 4 \log a + 2 \log b - 3 \log c$
(using law 3)
$= 4x + 2y - 3z$

5. $\log \sqrt{\dfrac{a}{c}}$

You must first rewrite the square root in index form:

$$\log \sqrt{\dfrac{a}{c}} = \log \left(\dfrac{a}{c}\right)^{\frac{1}{2}}$$

Then
$$\log \left(\dfrac{a}{c}\right)^{\frac{1}{2}} = \dfrac{1}{2}\log\left(\dfrac{a}{c}\right) \text{ (using law 3)}$$

$$= \dfrac{1}{2}(\log a - \log c) \text{ (using law 2)}$$

$$= \dfrac{1}{2}(x - z) \text{ or } \dfrac{1}{2}x - \dfrac{1}{2}z$$

Exercise 12B: *Write the following in terms x, y and/or z where $\log a = x$, $\log b = y$ and $\log c = z$.*

1. $\log ab$

2. $\log \dfrac{a}{b}$

3. $\log a^2 b$

4. $\log \dfrac{b}{c^3}$

5. $\log \sqrt{\dfrac{a}{b}}$

6. $\log \sqrt{\dfrac{b}{c}}$

7. $\log \dfrac{a^3 b}{c}$

8. $\log \dfrac{ab^4}{c^5}$

9. $\log \sqrt{\dfrac{a}{c^3}}$

10. $\log \sqrt[4]{\dfrac{a^2 c}{b}}$

12.3 Logarithms For Any Base

The general definition of a logarithm for any base is:
If $a^x = N$, then $x = \log_a N$

You need to know how to solve general log equations. The rules are as follows:
- Use the second part of the general definition to find the values of two of the unknowns.
- Use the first part of the general definition to find the missing value.

For example:

Find the value of a if (i) $\log_a 49 = 2$ (ii) $\log_a 0.2 = -1$
(i) The second part of the general definition states that:

$$\log_a N = x$$
Since $\log_a 49 = 2$
So $N = 49$, $x = 2$ and $a = a$

The first part of the general definition states that:
$$a^x = N$$
Therefore $a^2 = 49$
$$a = \sqrt{49}$$
$$a = 7$$

(ii) The second part of the general definition states that:
$$\log_a N = x$$
Since $\log_a 0.2 = -1$
So $N = 0.2$, $x = -1$ and $a = a$

The first part of the general definition states that:
$$a^x = N$$
Therefore $a^{-1} = 0.2$
So $\dfrac{1}{a} = \dfrac{1}{5}$
$$a = 5$$

Exercise 12C: *Find the value of a in each case:*

1. $\log_a 8 = 3$
2. $\log_a 0.25 = 1$
3. $\log_a 36 = 2$
4. $\log_a 9 = 2$
5. $\log_a 625 = 4$
6. $\log_a 32 = 5$
7. $\log_a 27 = 3$
8. $\log_a 125 = 3$
9. $\log_a 16 = 4$
10. $\log_a 64 = 3$
11. $\log_a 128 = 7$
12. $\log_a 216 = 3$

You can use the same method to find the number, where the log and the base are known. The rules are the same:
- Use the second part of the general definition to find the values of two of the unknowns.
- Use the first part of the general definition to find the missing value.

For example:

Find b if (i) $\log_9 b = 2$ (ii) $\log_8 b = -3$
(i) The second part of the general definition states that:
$$\log_a N = x$$
Since $\log_9 b = 2$
So $a = 9$, $x = 2$ and $N = b$

The first part of the general definition states that:
$$a^x = N$$
Therefore $9^2 = b$
$$b = 81$$

(ii) The second part of the general definition states that:

$$\log_a N = x$$

Since $\log_8 b = -3$

So $a = 8, x = -3$ and $N = b$

The first part of the general definition states that:

$$a^x = N$$

Therefore $8^{-3} = b$

So $b = (\frac{1}{8})^3$

$$b = \frac{1}{512}$$

Exercise 12D: *Find the value of b in each case:*

1. $\log_3 b = 4$
2. $\log_2 b = 8$
3. $\log_5 b = 2$
4. $\log_7 b = 3$
5. $\log_3 b = 2$
6. $\log_2 b = 3$
7. $\log_4 b = 2$
8. $\log_6 b = 2$
9. $\log_5 b = 3$
10. $\log_4 b = 3$
11. $\log_5 b = -1$
12. $\log_2 b = 5$
13. $\log_6 b = 3$
14. $\log_3 b = -2$
15. $\log_4 b = \frac{3}{2}$

12.4 Working Out Logs Of Numbers To Any Base

For example:

Work out the value of $\log_3 81$
You know that $\log_a N = x$
and you want $\log_3 81$
So $a = 3$ and $N = 81$
Now use the first part of the general definition to get:

$$a^x = N$$

Substituting: $3^x = 81$
And so $x = 4$

Exercise 12E: *Work out the value of the following:*

1. $\log_5 25$
2. $\log_6 216$
3. $\log_2 16$
4. $\log_3 27$
5. $\log_2 32$
6. $\log_8 64$
7. $\log_3 9$
8. $\log_4 32$
9. $\log_6 36$
10. $\log_8 4$
11. $\log_2 8$

12.5 Writing Logs Of Numbers In Terms Of Other Logs Of Prime Numbers

For example:

If log 3 = p and log 5 = q express the following in terms of p and q:

1. log 45
2. log 0.6
3. log 81

The rules are as follows:
• Rewrite each number in terms of 3 and/or 5,
• Use the appropriate law of logs,
• Then replace logs with p or q as appropriate.

1. Write 45 as a product of prime factors, ie keep dividing 45 by 3 and then 5 until you get 1:

```
3|45
3|15
5| 5
   1
```

So: $45 = 3 \times 3 \times 5$
$45 = 3^2 \times 5$

So: $\log 45 = \log (3^2 \times 5)$
$= \log 3^2 + \log 5$ (using law 1)
$= 2 \log 3 + \log 5$ (using law 3)
$= 2p + q$

2. Since 0.6 is less than 1 you must use division to get 0.6:

ie $0.6 = \frac{3}{5}$

So $\log 0.6 = \log \frac{3}{5}$
$= \log 3 - \log 5$ (using law 2)
$= p - q$

3. Keep dividing 81 by 3 until you get 1:

```
3|81
3|27
3| 9
3| 3
   1
```

So: $81 = 3 \times 3 \times 3 \times 3$
$81 = 3^4$

So: $\log 81 = \log 3^4$
$= 4 \log 3$ (using law 3)
$= 4p$

Exercise 12F:

1. If log 3 = p and log 2 = q express the following in terms of p and q:

(i) log 6 (ii) log 1.5 (iii) log 8 (iv) log 72

2. If log 2 = p and log 7 = q express the following in terms of p and q:

(i) log 3.5 (ii) log 28 (iii) log 16 (iv) log 98

3. If log 3 = p and log 11 = q express the following in terms of p and q:

(i) log 33 (ii) log 99 (iii) log 27 (iv) log 121

4. If log 2 = p and log 3 = q express the following in terms of p and q:

(i) log 6 (ii) log 32 (iii) log 48 (iv) log 1.5

5. If log 2 = p and log 5 = q express the following in terms of p and q:

(i) log 10 (ii) log 2.5 (iii) log 100 (iv) log 125

6. If log 2 = p and log 11 = q express the following in terms of p and q:

(i) log 22 (ii) log 5.5 (iii) log 64 (iv) log 44

7. If log 7 = p and log 5 = q express the following in terms of p and q:

(i) log 35 (ii) log 1.4 (iii) log 175 (iv) log 343

12.6 Writing Logs of Numbers Given to Different Bases in terms of Logs of Prime Numbers

You can do the same thing for logs that are expressed in bases other than 10. For example:

If $\log_7 5 = p$ and $\log_7 2 = q$ express the following in terms of p and q:

(i) $\log_7 10$
(ii) $\log_7 35$
(iii) $\log_7 3.5$
(iv) $\log_7 140$

The rules are as follows:
• Rewrite each number in terms of 5, 2 and/or 7,
• Use the appropriate law of logs,
• Then replace $\log_7 5 = p$, $\log_7 2 = q$ and $\log_7 7 = 1$ as appropriate.

(i) $10 = 2 \times 5$
So $\log_7 10 = \log_7 (2 \times 5)$
$= \log_7 2 + \log_7 5$ (using law 1)
$= p + q$

(ii) $35 = 5 \times 7$
So $\log_7 35 = \log_7 (5 \times 7)$
$= \log_7 5 + \log_7 7$ (using law 1)
$= p + 1$

(iii) $3.5 = \dfrac{7}{2}$
So $\log_7 3.5 = \log_7 \dfrac{3}{2}$
$= \log_7 7 - \log_7 2$ (using law 2)
$= 1 - q$

(iv)
$$\begin{array}{r|l} 2 & 140 \\ 2 & 70 \\ 5 & 35 \\ 7 & 7 \\ & 1 \end{array}$$

So: $140 = 2 \times 2 \times 5 \times 7$
$140 = 2^2 \times 5 \times 7$
So: $\log_7 140 = \log_7 (2^2 \times 5 \times 7)$
$= \log_7 2^2 + \log_7 5 + \log_7 7$ (using law 1)
$= 2\log_7 2 + \log_7 5 + \log_7 7$ (using law 3)
$= 2q + p + 1$

Exercise 12G:

1. If $\log_3 2 = p$ and $\log_3 5 = q$ express the following in terms of p and q:

(i) $\log_3 10$
(ii) $\log_3 20$
(iii) $\log_3 50$
(iv) $\log_3 6$
(v) $\log_3 45$
(vi) $\log_3 2.5$

2. If $\log_2 3 = p$ and $\log_2 7 = q$ express the following in terms of p and q:

(i) $\log_2 21$
(ii) $\log_2 63$
(iii) $\log_2 147$
(iv) $\log_2 24$
(v) $\log_2 28$

3. If $\log_5 2 = p$ and $\log_5 7 = q$ express the following in terms of p and q:

(i) $\log_5 3.5$
(ii) $\log_5 28$
(iii) $\log_5 98$
(iv) $\log_5 10$
(v) $\log_5 175$
(vi) $\log_5 70$

Exercise 12G...

4. If $\log_4 5 = p$ and $\log_4 7 = q$ express the following in terms of p and q:

      (i) $\log_4 35$
      (ii) $\log_4 1.4$
      (iii) $\log_4 175$
      (iv) $\log_4 20$
      (v) $\log_4 112$

5. If $\log_6 2 = p$ and $\log_6 11 = q$ express the following in terms of p and q:

      (i) $\log_6 22$
      (ii) $\log_6 5.5$
      (iii) $\log_6 12$
      (iv) $\log_6 44$
      (v) $\log_6 242$
      (vi) $\log_6 132$

6. If $\log_7 3 = p$ and $\log_7 5 = q$ express the following in terms of p and q:

      (i) $\log_7 15$
      (ii) $\log_7 21$
      (iii) $\log_7 35$
      (iv) $\log_7 45$
      (v) $\log_7 0.6$

7. If $\log_9 7 = p$ and $\log_9 11 = q$ express the following in terms of p and q:

      (i) $\log_9 77$
      (ii) $\log_9 63$
      (iii) $\log_9 99$
      (iv) $\log_9 539$
      (v) $\log_9 \dfrac{9}{49}$

8. If $\log_3 2 = p$ and $\log_3 11 = q$ express the following in terms of p and q:

      (i) $\log_3 6$
      (ii) $\log_3 99$
      (iii) $\log_3 5.5$
      (iv) $\log_3 44$
      (v) $\log_3 2.75$
      (vi) $\log_3 66$

9. If $\log_3 5 = p$ and $\log_3 7 = q$ express the following in terms of p and q:

      (i) $\log_3 35$
      (ii) $\log_3 45$
      (iii) $\log_3 21$
      (iv) $\log_3 1.4$
      (v) $\log_3 105$

CHAPTER 13: SOLVING INDEX EQUATIONS USING LOGARITHMS

13.1 Solving Index Equations

You can use law 3 of logarithms to solve index equations. The rules are as follows:
• Take logs of both sides,
• Use law 3 to change the equation into a linear equation,
• Solve the linear equation.

For example:

Solve (i) $7^x = 45$ (ii) $8^{-x} = 1.96$

(i)
$$7^x = 45$$
$$\log 7^x = \log 45$$
$$x \log 7 = \log 45 \text{ (using law 3)}$$
$$x = \frac{\log 45}{\log 7}$$
$$x = 1.956...$$
$$x = 1.96 \text{ to 3 significant figures}$$

(ii)
$$8^{-x} = 1.96$$
$$\log 8^{-x} = \log 1.96$$
$$-x \log 8 = \log 1.96 \text{ (using law 3)}$$
$$x = \frac{-\log 1.96}{\log 8}$$
$$x = -0.3236...$$
$$x = -0.324 \text{ to 3 significant figures}$$

Exercise 13A: *Solve the following, giving your answers correct to 3 significant figures:*

1. $4^x = 7$	**7.** $8^x = 20$
2. $7^{-x} = 5$	**8.** $5^{-x} = 30$
3. $5^x = 3$	**9.** $3^x = 16$
4. $9^x = 4$	**10.** $2^x = 35$
5. $4^{-x} = 10$	**11.** $3^{-x} = 8$
6. $6^x = 15$	**12.** $12^x = 5$

13.2 More Complex Index Equations

You can use the same method for more complex indices. The rules are the same:

- Take logs of both sides,
- Use law 3 to change the equation into a linear equation,
- Solve the linear equation.

For example:

Solve (i) $6^{2x-1} = 5$ (ii) $4^{1-\frac{1}{3}x} = 11$

(i)
$$6^{2x-1} = 5$$
$$\log 6^{2x-1} = \log 5$$
$$(2x-1)\log 6 = \log 5 \text{ (using law 3)}$$

Note: You must include the bracket around $2x-1$ as there are two algebraic terms.

$$2x - 1 = \frac{\log 5}{\log 6}$$
$$2x - 1 = 0.8982$$
$$2x = 0.8982 + 1$$
$$2x = 1.8982$$
$$x = 0.9491$$
$$x = 0.949 \text{ to 3 significant figures}$$

(ii)
$$4^{1-\frac{1}{3}x} = 11$$
$$\log 4^{1-\frac{1}{3}x} = \log 11$$
$$(1 - \frac{1}{3}x)\log 4 = \log 11 \quad \text{(using law 3)}$$

Note: You must include the bracket around $1 - \frac{1}{3}x$ as there are 2 algebraic terms.

$$1 - \frac{1}{3}x = \frac{\log 11}{\log 4}$$
$$1 - \frac{1}{3}x = 1.7297$$
$$-\frac{1}{3}x = 1.7297 - 1$$
$$-\frac{1}{3}x = 0.7297$$

Multiply both sides by -3 to get:
$$x = -2.1891$$
$$x = -2.19 \text{ to 3 significant figures}$$

1. $7^{3x+2} = 9$
2. $4^{5x-2} = 6$
3. $2^{1-2x} = 5$
4. $3^{4-\frac{1}{2}x} = 8$
5. $9^{3x} = 7$
6. $6^{5+\frac{1}{4}x} = 8$
7. $3^{4x-1} = 12$
8. $5^{2-5x} = 9$
9. $4^{3x+2} = 15$
10. $2^{1+\frac{3}{4}x} = 9$

13.3 Further Index Equations

You can use the same method when each side of the equation has an index with the same unknown. The rules are the same:

- Take logs of both sides,
- Use law 3 to change the equation into a linear equation,
- Solve the linear equation.

For example:

Solve (i) $4^{3x-1} = 7^{x+2}$ (ii) $4^{1-2x} = 3^{4x-1}$

(i)
$$4^{3x-1} = 7^{x+2}$$
$$\log 4^{3x-1} = \log 7^{x+2}$$
$$(3x-1)\log 4 = (x+2)\log 7 \text{ (using law 3)}$$

Note: You must include both brackets as there are two algebraic terms on each side.
Work out each bracket by multiplying to get:

$$3x\log 4 - \log 4 = x\log 7 + 2\log 7$$
$$3x\log 4 - x\log 7 = 2\log 7 + \log 4$$
$$x(3\log 4 - \log 7) = 2\log 7 + \log 4$$
$$0.961x = 2.292$$
$$x = \frac{2.292}{0.961}$$
$$x = 2.385$$
$$x = 2.39 \text{ to 3 significant figures}$$

(ii)
$$4^{1-2x} = 3^{4x-1}$$
$$\log 4^{1-2x} = \log 3^{4x-1}$$
$$(1-2x)\log 4 = (4x-1)\log 3 \text{ (using law 3)}$$

Note: You must include both brackets as there are two algebraic terms on each side.
Work out each bracket by multiplying to get:

$$\log 4 - 2x\log 4 = 4x\log 3 - \log 3$$
$$\log 4 + \log 3 = 4x\log 3 + 2x\log 4$$
$$\log 4 + \log 3 = x(4\log 3 + 2\log 4)$$
$$1.079 = 3.113x$$
$$x = \frac{1.079}{3.113}$$
$$x = 0.3466$$
$$x = 0.347 \text{ to 3 significant figures}$$

1. $5^{2x-1} = 3^{x+4}$
2. $4^{3x+2} = 7^{x-5}$
3. $2^{1-x} = 5^{x+2}$
4. $5^{1-3x} = 8^{4-x}$
5. $9^{3x+2} = 7^{1-x}$
6. $4^{x-2} = 3^{4-x}$
7. $6^{2x-1} = 3^{x+4}$
8. $6^{5x-2} = 2^{3x+5}$
9. $4^{1-3x} = 3^{x+2}$
10. $6^{2x+5} = 2^{5x-2}$

CHAPTER 14: VECTORS

14.1 Translations

The diagram shows four different translations from one point to another point.

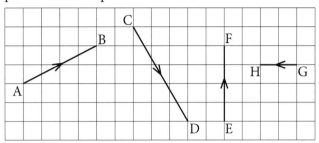

Each translation is defined by a **vector**.

A vector has **two** components: $\begin{pmatrix} x \\ y \end{pmatrix}$

The x component shows how much the point moves across:
• It is **positive** if the point moves to the **right**.
• It is **negative** if the point moves to the **left**.
The y component shows how much the point moves across:
• It is **positive** if the point moves **up**.
• It is **negative** if the point moves **down**.

The vector which defines the translation from A to B is labelled $\overrightarrow{AB}$. Similarly the vectors which define the translations from C to D, E to F and G to H are labelled $\overrightarrow{CD}$, $\overrightarrow{EF}$ and $\overrightarrow{GH}$.

The translation from A to B is:
'4 to the right and then 2 up.'
You write the vector $\overrightarrow{AB}$ as:

$\overrightarrow{AB} = \begin{pmatrix} 4 \\ 2 \end{pmatrix}$

The translation from C to D is:
'3 to the right and then 5 down.'
You write the vector $\overrightarrow{CD}$ as:

$\overrightarrow{CD} = \begin{pmatrix} 3 \\ -5 \end{pmatrix}$

The translation from E to F is '4 up.'
You write the vector $\overrightarrow{EF}$ as:

$\overrightarrow{EF} = \begin{pmatrix} 0 \\ 4 \end{pmatrix}$

The translation from G to H is '2 to the left.'
You write the vector $\overrightarrow{GH}$ as:

$\overrightarrow{GH} = \begin{pmatrix} -2 \\ 0 \end{pmatrix}$

For example:

Show the following vectors:

(i) $\begin{pmatrix} 2 \\ -5 \end{pmatrix}$ (ii) $\begin{pmatrix} -3 \\ 4 \end{pmatrix}$ (iii) $\begin{pmatrix} 2 \\ 0 \end{pmatrix}$ (iv) $\begin{pmatrix} 0 \\ -3 \end{pmatrix}$

(i) It doesn't matter where you start. But you must show the correct direction with an arrow.

(i) $\begin{pmatrix} 2 \\ -5 \end{pmatrix}$ means '2 to the right and 5 down'.

(ii) $\begin{pmatrix} -3 \\ 4 \end{pmatrix}$ means '3 to the left and 4 up'.

(iii) $\begin{pmatrix} 2 \\ 0 \end{pmatrix}$ means '2 to the right'.

(iv) $\begin{pmatrix} 0 \\ -3 \end{pmatrix}$ means '3 down'.

You can draw each one as follows:

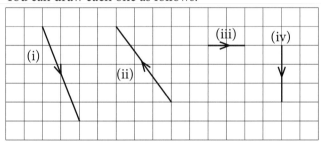

Exercise 14A: Show the vectors below on a squared grid:

1. $\begin{pmatrix} 3 \\ -2 \end{pmatrix}$

2. $\begin{pmatrix} -2 \\ 4 \end{pmatrix}$

3. $\begin{pmatrix} 4 \\ 7 \end{pmatrix}$

4. $\begin{pmatrix} -1 \\ 3 \end{pmatrix}$

5. $\begin{pmatrix} -2 \\ -4 \end{pmatrix}$

Write down the following vectors in component form:

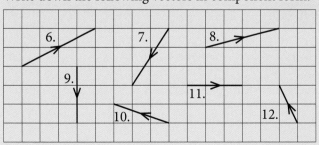

14.2 Vector Notation

Vectors can also be defined by small letters. For example, the vectors below are called **a** and **b**.

14.3 Adding And Subtracting Vectors

You must **add** vectors by starting the second vector at the end of the first vector. This is called adding vectors '**nose–to–tail**'. For example if **a** and **b** are vectors:

then **a** + **b** is given by:

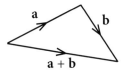

You must **subtract** vectors by adding the negative of the second vector to the end of the first vector. This is because **a** – **b** is the same as **a** + (–**b**). For example, if **P** and **Q** are vectors:

then **P** – **Q** is given by:

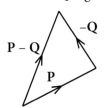

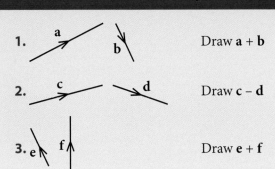

1. Draw **a** + **b**

2. Draw **c** – **d**

3. Draw **e** + **f**

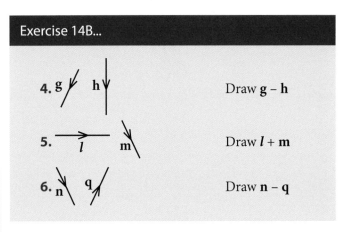

4. Draw **g** – **h**

5. Draw *l* + **m**

6. Draw **n** – **q**

14.4 Multiplying A Vector By A Number

If you multiply a vector **a** by a **positive** number k then the vector k**a** is a vector in the **same** direction as **a** but k times as long.

If you multiply a vector **a** by a **negative** number k then the vector k**a** is a vector in the **opposite** direction as **a** but k times as long.

For example if **a** is a vector:

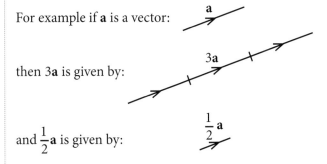

then 3**a** is given by:

and $\frac{1}{2}$**a** is given by:

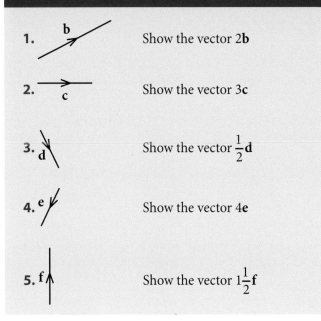

1. Show the vector 2**b**

2. Show the vector 3**c**

3. Show the vector $\frac{1}{2}$**d**

4. Show the vector 4**e**

5. Show the vector $1\frac{1}{2}$**f**

Exercise 14C...

6. 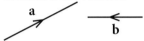 Show the vector $-2\mathbf{g}$

7. h Show the vector $-\frac{1}{2}\mathbf{h}$

8. p Draw $-\mathbf{p}$

You can and and subtract vectors that have been multiplied or divided by a number. For example:

If **a** and **b** are vectors:

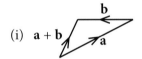

Show the vectors:
(i) $\mathbf{a} + \mathbf{b}$
(ii) $\mathbf{a} - \mathbf{b}$
(iii) $2\mathbf{a} + \mathbf{b}$
(iv) $\mathbf{b} - 3\mathbf{a}$

(i) $\mathbf{a} + \mathbf{b}$

(ii)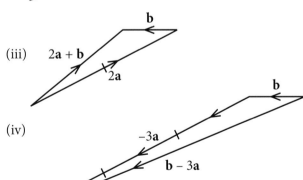

(iii) $2\mathbf{a} + \mathbf{b}$

(iv)

Exercise 14D: *Show the following vectors. You don't need to draw them exactly to scale.*

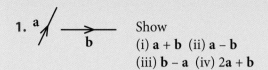

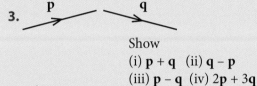

1. a b Show
(i) $\mathbf{a} + \mathbf{b}$ (ii) $\mathbf{a} - \mathbf{b}$
(iii) $\mathbf{b} - \mathbf{a}$ (iv) $2\mathbf{a} + \mathbf{b}$

2. c d Show
(i) $\mathbf{c} + \mathbf{d}$ (ii) $\mathbf{d} - \mathbf{c}$
(iii) $\mathbf{c} - \mathbf{d}$ (iv) $\mathbf{d} - 2\mathbf{c}$

3. p q Show
(i) $\mathbf{p} + \mathbf{q}$ (ii) $\mathbf{q} - \mathbf{p}$
(iii) $\mathbf{p} - \mathbf{q}$ (iv) $2\mathbf{p} + 3\mathbf{q}$

4. c d Show
(i) $\mathbf{c} - \mathbf{d}$ (ii) $\mathbf{d} + \mathbf{c}$
(iii) $\mathbf{d} - \mathbf{c}$ (iv) $2\mathbf{d} - 3\mathbf{c}$

5. a b Show
(i) $\mathbf{a} + \mathbf{b}$ (ii) $\mathbf{a} - \mathbf{b}$
(iii) $\mathbf{b} - \mathbf{a}$ (iv) $2\mathbf{a} + \mathbf{b}$

6. x y Show
(i) $\mathbf{x} - \mathbf{y}$ (ii) $\mathbf{y} + \mathbf{x}$
(iii) $\mathbf{y} - \mathbf{x}$ (iv) $2\mathbf{y} + \mathbf{x}$

7. a b Show
(i) $\mathbf{a} - \mathbf{b}$ (ii) $\mathbf{a} + \mathbf{b}$
(iii) $\mathbf{b} - \mathbf{a}$ (iv) $2\mathbf{b} + 3\mathbf{a}$

8. p q Show
(i) $\mathbf{p} - \mathbf{q}$ (ii) $\mathbf{q} - \mathbf{p}$
(iii) $\mathbf{p} + \mathbf{q}$ (iv) $2\mathbf{p} + 3\mathbf{q}$

9. x y Show
(i) $\mathbf{x} + \mathbf{y}$ (ii) $\mathbf{x} - \mathbf{y}$
(iii) $\mathbf{y} - \mathbf{x}$ (iv) $2\mathbf{x} + 3\mathbf{y}$

10. p q Show
(i) $\mathbf{p} - \mathbf{q}$ (ii) $\mathbf{q} + \mathbf{p}$
(iii) $\mathbf{q} - \mathbf{p}$ (iv) $2\mathbf{p} - 3\mathbf{q}$

14.5 Working Out Different Vectors

For example:

1. The vectors $\overrightarrow{PQ} = \mathbf{v}$ and $\overrightarrow{PR} = 2\mathbf{n}$ are shown below:

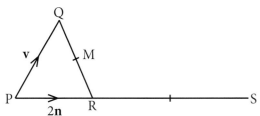

M is the midpoint of QR and S is the point on PR **produced** so that RS = 2PR (note: to **produce** a line PR means to extend the line PR in the direction PR). Find in terms of $\mathbf{v}$ and/or $\mathbf{n}$ the vectors:
(i) $\overrightarrow{PM}$ (ii) $\overrightarrow{PS}$ (iii) $\overrightarrow{SM}$

(i) $\overrightarrow{PM} = \overrightarrow{PQ} + \overrightarrow{QM}$

$\overrightarrow{PM} = \overrightarrow{PQ} + \frac{1}{2}\overrightarrow{QR}$

But $\overrightarrow{QR} = \overrightarrow{QP} + \overrightarrow{PR}$

$\overrightarrow{QR} = -\mathbf{v} + 2\mathbf{n}$

So $\overrightarrow{PM} = \mathbf{v} + \frac{1}{2}(-\mathbf{v} + 2\mathbf{n})$

$\overrightarrow{PM} = \mathbf{v} - \frac{1}{2}\mathbf{v} + \mathbf{n}$

$\overrightarrow{PM} = \frac{1}{2}\mathbf{v} + \mathbf{n}$

(ii) $\overrightarrow{PS} = \overrightarrow{PR} + \overrightarrow{RS}$

$\overrightarrow{PS} = 2\mathbf{n} + 2(2\mathbf{n})$

$\overrightarrow{PS} = 2\mathbf{n} + 4\mathbf{n}$

$\overrightarrow{PS} = 6\mathbf{n}$

(iii) $\overrightarrow{SM} = \overrightarrow{SR} + \overrightarrow{RM}$

$\overrightarrow{SM} = -4\mathbf{n} + \frac{1}{2}(\mathbf{v} - 2\mathbf{n})$

$\overrightarrow{SM} = -4\mathbf{n} + \frac{1}{2}\mathbf{v} - \mathbf{n}$

$\overrightarrow{SM} = -5\mathbf{n} + \frac{1}{2}\mathbf{v}$

2. PQRS is a trapezium as shown below, in which SR and PQ are parallel.

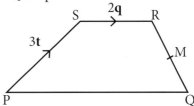

$PQ = \frac{5}{2} SR$ and M is the midpoint of RQ.

The vectors $\overrightarrow{PS}$ and $\overrightarrow{SR}$ are defined by $3\mathbf{t}$ and $2\mathbf{q}$ respectively. Find these vectors in terms of $\mathbf{t}$ and/or $\mathbf{q}$:
(i) $\overrightarrow{PQ}$ (ii) $\overrightarrow{RQ}$ (iii) $\overrightarrow{PM}$

(i) $\overrightarrow{PQ} = \frac{5}{2} \overrightarrow{SR}$

$\overrightarrow{PQ} = \frac{5}{2} \times 2\mathbf{q}$

$\overrightarrow{PQ} = 5\mathbf{q}$

(ii) $\overrightarrow{RQ} = \overrightarrow{RS} + \overrightarrow{SP} + \overrightarrow{PQ}$

$\overrightarrow{RQ} = -2\mathbf{q} - 3\mathbf{t} + 5\mathbf{q}$ using answer to (i)

$\overrightarrow{RQ} = 3\mathbf{q} - 3\mathbf{t}$

(iii) $\overrightarrow{PM} = \overrightarrow{PQ} + \overrightarrow{QM}$

$\overrightarrow{PM} = \overrightarrow{PQ} + \frac{1}{2}\overrightarrow{QR}$

$\overrightarrow{PM} = 5\mathbf{q} + \frac{1}{2}(-3\mathbf{q} + 3\mathbf{t})$

$\overrightarrow{PM} = 5\mathbf{q} - \frac{3}{2}\mathbf{q} + \frac{3}{2}\mathbf{t}$

$\overrightarrow{PM} = \frac{7}{2}\mathbf{q} + \frac{3}{2}\mathbf{t}$

Exercise 14E:

1. ABC is a triangle. M is the midpoint of AC. The vectors $\overrightarrow{BA}$ and $\overrightarrow{BC}$ are defined by $\mathbf{x}$ and $\mathbf{y}$ respectively.

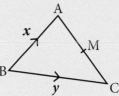

(a) Find in terms of $\mathbf{x}$ and/or $\mathbf{y}$ the vectors
(i) $\overrightarrow{AC}$ (ii) $\overrightarrow{BM}$

D is the point on BC produced so that BC = 2CD

(b) Find in terms of $\mathbf{x}$ and/or $\mathbf{y}$ the vectors
(i) $\overrightarrow{BD}$ (ii) $\overrightarrow{AD}$

Exercise 14E:

2. DEF is a triangle. M is the midpoint of EF. The vectors $\overrightarrow{ED}$ and $\overrightarrow{DF}$ are defined by **2a** and **3b** respectively.

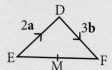

(a) Find in terms of **a** and/or **b** the vectors
(i) $\overrightarrow{EF}$ (ii) $\overrightarrow{DM}$

G is the point on DF so that DG:GF = 1:3

(b) Find in terms of **a** and/or **b** the vectors
(i) $\overrightarrow{DG}$ (ii) $\overrightarrow{MG}$

3. LMNP is a parallelogram. The vectors $\overrightarrow{PL}$ and $\overrightarrow{PN}$ are defined by **a** and **3b** respectively.

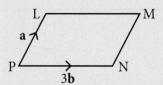

(a) Find in terms of **a** and/or **b** the vectors
(i) $\overrightarrow{PM}$ (ii) $\overrightarrow{NL}$

T is the point on NM produced so that MT = 2NM.

(b) Find in terms of **a** and/or **b** the vectors
(i) $\overrightarrow{NT}$ (ii) $\overrightarrow{PT}$

4. ABCD is a rectangle. M is the midpoint of BC. The vectors $\overrightarrow{DA}$ and $\overrightarrow{DB}$ are defined by **a** and **b** respectively.

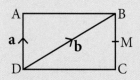

(a) Find in terms of **a** and/or **b** the vectors
(i) $\overrightarrow{AB}$ (ii) $\overrightarrow{DM}$

X is the point on DC so that $DX = \frac{2}{3}CD$

(b) Find in terms of **a** and/or **b** the vectors
(i) $\overrightarrow{DX}$ (ii) $\overrightarrow{MX}$

5. ABCDEF is a regular hexagon. The vectors $\overrightarrow{BA}$ and $\overrightarrow{AF}$ are defined by **3q** and **2p** respectively.

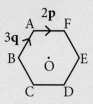

Find in terms of **p** and/or **q** the vectors
(i) $\overrightarrow{BE}$ (ii) $\overrightarrow{FE}$ (iii) $\overrightarrow{DF}$ (iv) $\overrightarrow{AD}$

6. PQR is a triangle. M is the midpoint of PR. The vectors $\overrightarrow{QP}$ and $\overrightarrow{QR}$ are defined by **4x** and **3y** respectively.

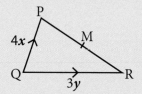

(a) Find in terms of **x** and/or **y** the vectors

(i) $\overrightarrow{PR}$ (ii) $\overrightarrow{QM}$

T is the point on QR produced so that RT = 2QR.

(b) Find in terms of **x** and/or **y** the vectors

(i) $\overrightarrow{QT}$ (ii) $\overrightarrow{TM}$

7. ABCD is a parallelogram. M is the midpoint of BC. The vectors $\overrightarrow{AC}$ and $\overrightarrow{DC}$ are defined by **b** and **a** respectively.

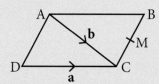

(a) Find in terms of **a** and/or **b** the vectors

(i) $\overrightarrow{AD}$ (ii) $\overrightarrow{AM}$

T is the point on DC so that DT:TC = 2:3

(b) Find in terms of **a** and/or **b** the vectors (i) $\overrightarrow{DT}$
(ii) $\overrightarrow{MT}$

Exercise 14E...

8. ABCD is a trapezium in which AB and DC are parallel. DC is $\frac{4}{3}$ AB. M is the midpoint of BC.

The vectors $\overrightarrow{DA}$ and $\overrightarrow{AB}$ are defined by $2x$ and $3y$ respectively.

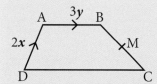

(a) Find in terms of x and/or y the vectors

(i) $\overrightarrow{DC}$ (ii) $\overrightarrow{BC}$ (iii) $\overrightarrow{DM}$

Q is the point on DC produced so that DQ = 3DC

(b) Find in terms of x and/or y the vectors

(i) $\overrightarrow{DQ}$ (ii) $\overrightarrow{MQ}$

9. ABCDEF is a regular hexagon. The vectors $\overrightarrow{BA}$ and $\overrightarrow{BE}$ are defined by $2b$ and $4a$ respectively.

Find in terms of a and/or b the vectors

(i) $\overrightarrow{AF}$ (ii) $\overrightarrow{FE}$ (iii) $\overrightarrow{AC}$ (iv) $\overrightarrow{CF}$

10. ABCD is a trapezium in which AB and DC are parallel. AB = $\frac{2}{3}$ DC. M is the midpoint of AB. The vectors $\overrightarrow{DA}$ and $\overrightarrow{DC}$ are defined by a and $3b$ respectively.

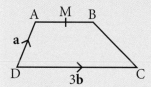

(a) Find in terms of a and/or b the vectors

(i) $\overrightarrow{AB}$ (ii) $\overrightarrow{CB}$ (iii) $\overrightarrow{DM}$

Q is the point on DC so that DQ:DC = 1:3

(b) Find in terms of a and/or b the vectors

(i) $\overrightarrow{DQ}$ (ii) $\overrightarrow{MQ}$

14.6 Making Deductions Based On Vectors

You can always make two deductions, one based on the magnitude of the vectors and one based on the direction of the vectors.

If $\overrightarrow{AB} = k\overrightarrow{CD}$ then you can deduce that:

- AB is k times as long as CD;
- AB is parallel to CD.

If $\overrightarrow{AB} = k\overrightarrow{AC}$ then you can deduce that:

- AB is k times as long as AC;
- A, B and C all lie on a straight line (ie A, B and C are collinear).

For example:

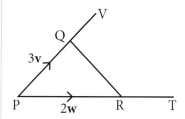

PQR is a triangle. The vectors $\overrightarrow{PQ}$ and $\overrightarrow{PR}$ are defined by $3v$ and $2w$ respectively.

T is the point on PR produced so that RT = $\frac{1}{2}$PR

V is the point on PQ produced so that PQ:QV = 2:1

(a) Find in terms of v and/or w the vectors
(i) $\overrightarrow{QR}$ (ii) $\overrightarrow{VT}$

(b) What deductions can you make about QR and VT?

(a)(i) $$\overrightarrow{QR} = \overrightarrow{QP} + \overrightarrow{PR}$$
$$\overrightarrow{QR} = -3v + 2w$$

(ii) Since PQ:QV = 2:1

So $$\overrightarrow{QV} = 1\frac{1}{2}v$$

Therefore $$\overrightarrow{PV} = 4\frac{1}{2}v$$

And since RT = $\frac{1}{2}$ PR

So $$\overrightarrow{RT} = \frac{1}{2}\,2w = w$$

Therefore $$\overrightarrow{PT} = 2w + w = 3w$$

And since $$\overrightarrow{VT} = \overrightarrow{VP} + \overrightarrow{PT}$$

Therefore $$\overrightarrow{VT} = -4\frac{1}{2}v + 3w$$

(b) Note that $\overrightarrow{QR} = -3\mathbf{v} + 2\mathbf{w}$

Looking at $\overrightarrow{VT} = -4\frac{1}{2}\mathbf{v} + 3\mathbf{w}$

We can say $\overrightarrow{VT} = 1\frac{1}{2}(-3\mathbf{v} + 2\mathbf{w})$

(Since $-4\frac{1}{2} = 1\frac{1}{2} \times (-3)$ and $3 = 1\frac{1}{2} \times 2$)

Thus $\overrightarrow{VT} = 1\frac{1}{2}\overrightarrow{QR}$

So we can deduce that:

- VT is $1\frac{1}{2}$ times as long as QR.
- VT and QR are parallel.

Exercise 14F:

1. ABCD is a parallelogram. The vectors $\overrightarrow{AB}$ and $\overrightarrow{AD}$ are defined by **2p** and **3q** respectively.

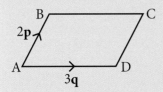

(a) Find in terms of **p** and/or **q** the vectors
(i) $\overrightarrow{AC}$ (ii) $\overrightarrow{BD}$

(b) M is the midpoint of BD. Prove by vectors that A, M and C are collinear.

2. ABCD is a rectangle. The vectors $\overrightarrow{AC}$ and $\overrightarrow{AD}$ are defined by **x** and **y** respectively

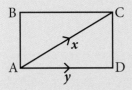

(a) Find $\overrightarrow{DC}$ in terms of **x** and **y**.
E is the point on AD produced so that AD:DE = 1:2
F is the point on AC produced so that AF = 3AC
(b) Prove by vectors that EF is parallel to DC.

3. ABCD is a trapezium in which AD and BC are parallel.

$AD = \frac{4}{3}BC$. M is the midpoint of AD.

N is the point on AB produced so that AB = BN.

The vectors $\overrightarrow{AB}$ and $\overrightarrow{AD}$ are defined by **2p** and **8q** respectively.

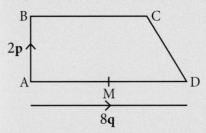

(a) Find in terms of **p** and/or **q** the vectors
(i) $\overrightarrow{BC}$ (ii) $\overrightarrow{DC}$ (iii) $\overrightarrow{MN}$

(b) What deductions can you make about MN and DC?

4. PQR is a triangle. U is the point on PR so that PU:UR = 1:2. T is the point on PQ so that $PT = \frac{1}{3}PQ$.

The vectors $\overrightarrow{PR}$ and $\overrightarrow{PT}$ are defined by **4b** and **a** respectively.

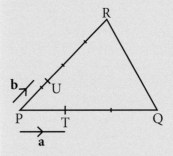

(a) Find in terms of **a** and/or **b** the vectors
(i) $\overrightarrow{PQ}$ (ii) $\overrightarrow{QR}$ (iii) $\overrightarrow{UT}$

(b) What deductions can you make about QR and TU?

5. ABCDEF is a regular hexagon. The vectors $\overrightarrow{FE}$ and $\overrightarrow{FO}$ are defined by **3t** and **2v** respectively. G is the point on FA produced so that AG = 2FA.

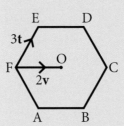

(a) Find in terms of **t** and/or **v** the vectors
(i) $\overrightarrow{EO}$ (ii) $\overrightarrow{DO}$ (iii) $\overrightarrow{DB}$ (iv) $\overrightarrow{BG}$

(b) What deductions can you make about DB and BG?

15.1 Displacement/Time Graphs

This type of graph is formed when you plot **time** along the horizontal axis and **displacement** along the vertical axis.

The **gradient** of each section of the displacement/time graph tells you the **velocity** in that section. You can work out the gradient using:

$$\text{Gradient} = \frac{\text{Rise}}{\text{Run}}$$

A horizontal line segment indicates a velocity of 0 m/s (ie the body has stopped). For example:

Work out the velocities in each section of the displacement/time graph below.

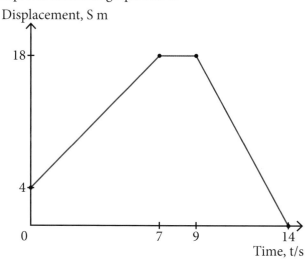

In the first section the gradient is given by:

$$\text{Gradient} = \frac{18 - 4}{7}$$

$$\text{So velocity} = \frac{14}{7} = 2 \text{ m/s}$$

In the middle section the line is horizontal:

$$\text{So velocity} = 0 \text{ m/s}$$

In the final section the gradient is given by:

$$\text{Gradient} = \frac{0 - 18}{5}$$

$$\text{So velocity} = \frac{-18}{5} = -3.6$$

which means 3.6 m/s in the opposite direction.

Exercise 15A: *Work out the velocities in each section of the displacement/time graphs below.*

1.

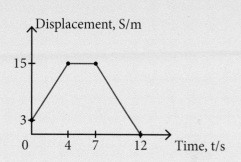

2.

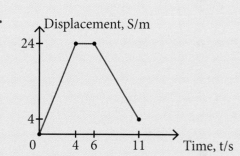

3.

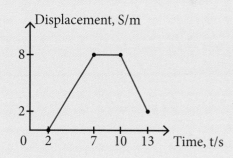

4.

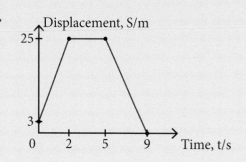

Exercise 15A...

5.

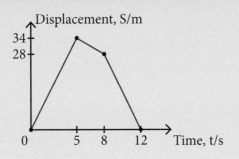

6.

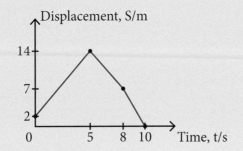

15.2 Velocity/Time Graphs

This type of graph is formed when you plot **time** along the horizontal axis and **velocity** along the vertical axis. The **gradient** of each section of the velocity/time graph tells you the **acceleration** in that section. You can work out the gradient using:

$$\text{Gradient} = \frac{\text{Rise}}{\text{Run}}$$

A horizontal line segment indicates an acceleration of 0 m/s^2 (ie the body is moving at a constant velocity). The **area under** the velocity/time graph tells you the **distance travelled** during the motion.
For example:

In the velocity/time graph below, work out the
(i) acceleration in each section
(ii) total distance travelled.

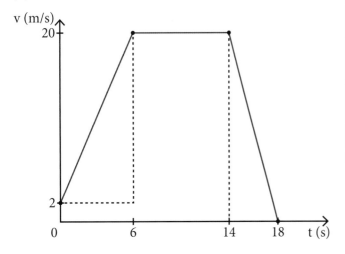

(i) In the first section the gradient is given by:

$$\text{Gradient} = \frac{20 - 2}{6}$$

$$\text{So acceleration} = \frac{18}{6} = 3 \text{ m/s}^2$$

In the middle section the line is horizontal:
So acceleration $= 0$ m/s^2

In the final section the gradient is given by:

$$\text{Gradient} = \frac{0 - 20}{4}$$

$$\text{So acceleration} = \frac{-20}{4} = -5 \text{ m/s}^2$$

ie, the body is slowing down, or decelerating.

Note: If the question had asked for the **deceleration** in the last part then the answer would be 5 m/s^2.

(ii) The total distance travelled is given by the area under the velocity/time graph. You must split the area under the graph into different regular shapes as shown below:

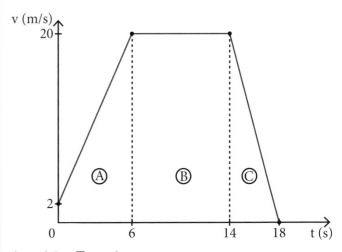

Area A is a Trapezium, so:

$$\text{Area} = \frac{1}{2}(2 + 20) \times 6$$

$$= \frac{1}{2} \times 22 \times 6$$

$$= 66$$

Area B is a rectangle, so:

$$\text{Area} = 8 \times 20$$

$$= 160$$

Area C is a triangle, so:

$$\text{Area} = \frac{1}{2} \times 4 \times 20$$

$$= 40$$

So total distance $= 66 + 160 + 40$

$$= 266 \text{ m}$$

Exercise 15B:

In the velocity/time graphs in questions 1 to 10, work out
(i) the acceleration in each section of the graph
(ii) the total distance travelled.

1. v (m/s)

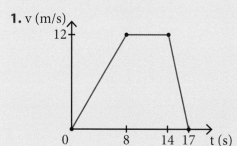

2. v (m/s)

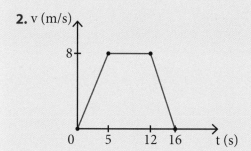

3. v (m/s)

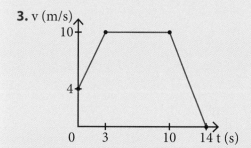

4. v (m/s)

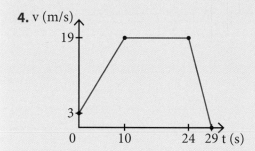

5. v (m/s)

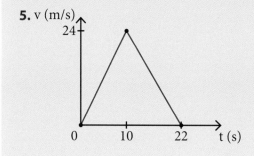

6. v (m/s)

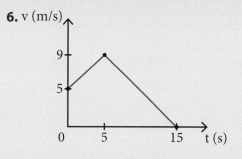

7. v (m/s)

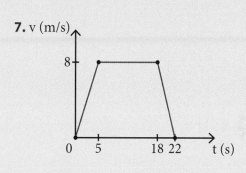

8. v (m/s)

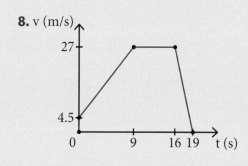

9. v (m/s)

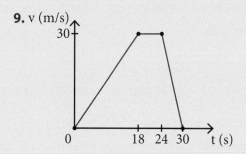

10. v (m/s)

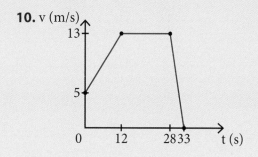

11. A body accelerates uniformly from 3 m/s to 8 m/s in 6 s. It then continues at this velocity for 10 s before decelerating to rest in a further 5 s.

(a) Draw a velocity/time graph to show this journey.

(b) Use your graph to find the

    (i) initial acceleration

    (ii) final deceleration

    (iii) total distance travelled.

12. A body accelerates uniformly from 8 m/s to 20 m/s in 6 s. It then continues at this velocity for 5 s before decelerating to rest in a further 10 s.

(a) Draw a velocity/time graph to show this journey.

(b) Use your graph to find the

    (i) initial acceleration

    (ii) final deceleration

    (iii) total distance travelled.

13. A body accelerates uniformly from 3 m/s to 11 m/s in 4 s. It then continues at this velocity for 6 s before decelerating to rest in a further 2 s.

(a) Draw a velocity/time graph to show this journey.

(b) Use your graph to find the

    (i) initial acceleration

    (ii) final deceleration

    (iii) total distance travelled.

14. A body accelerates uniformly from 2 m/s to 9 m/s in 5 s. It then continues at this velocity for 8 s before decelerating to rest in a further 3 s.

(a) Draw a velocity/time graph to show this journey.

(b) Use your graph to find the

    (i) initial acceleration

    (ii) final deceleration

    (iii) total distance travelled.

15. A body accelerates uniformly from rest to 8 m/s in 4 s. It then continues at this velocity for 7 s before decelerating to 2 m/s in a further 5 s.

(a) Draw a velocity/time graph to show this journey.

(b) Use your graph to find the

    (i) initial acceleration

    (ii) final deceleration

    (iii) total distance travelled.

16. A body accelerates uniformly from 5 m/s to 12 m/s in 4 s. It then continues at this velocity for 12 s before decelerating to rest in a further 6 s.

(a) Draw a velocity/time graph to show this journey.

(b) Use your graph to find the

    (i) initial acceleration

    (ii) final deceleration

    (iii) total distance travelled.

17. A body accelerates uniformly from rest to 20 m/s in 5 s. It then continues at this velocity for 8 s before decelerating to 4 m/s in a further 9 s

(a) Draw a velocity/time graph to show this journey.

(b) Use your graph to find the

    (i) initial acceleration

    (ii) final deceleration

    (iii) total distance travelled

15.3 Uniformly Accelerated Motion

Consider a body which accelerates uniformly from **u** m/s to **v** m/s in **t** seconds at an acceleration of **a** m/s² covering a distance of **s** metres. Its velocity/time graph is shown below:

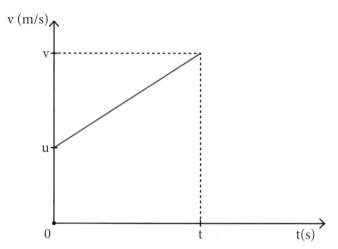

The acceleration is given by the gradient.

So $a = \dfrac{v - u}{t}$

$v - u = at$

So $\mathbf{v = u + at}$ **Equation 1**

The distance s is given by the area of the trapezium.

So $\mathbf{s = \dfrac{1}{2}\,t\,(u + v)}$ **Equation 2**

You can rearrange Equation 1 to get:

$u + at = v$

$at = v - u$

$t = \dfrac{v - u}{a}$

Substituting this into Equation 2 gives:

$s = \dfrac{1}{2}\dfrac{v - u}{a}(u + v)$

Multiplying each side by 2a gives:

$2as = (v - u)(u + v)$

$2as = v^2 - u^2$

So $\mathbf{v^2 = u^2 + 2as}$ **Equation 3**

Substituting v = u + at into Equation 2 gives:

$s = \dfrac{1}{2}\,t\,(u + v)$

$s = \dfrac{1}{2}\,t\,(u + u + at)$

$s = \dfrac{1}{2}\,t\,(2u + at)$

$\mathbf{s = ut + \dfrac{1}{2}at^2}$ **Equation 4**

These are the four **equations of motion**. They will be used below and in chapter 16.

15.4 Problems Involving Two Journeys

Sometimes you need to solve problems involving two journeys. For example:

Quinn sets off from home walking at a constant velocity of 1.8 m/s. Two minutes later Martin sets off from home cycling from rest at an acceleration of 2.4 m/s² until he reaches a velocity of 12 m/s. He then continues at this velocity until he catches up with Quinn. Martin then decelerates uniformly to rest in a further 4 s.

(a) Draw a velocity/time graph to show these journeys.

(b) (i) Find the time Quinn has walked when Martin catches up with him.

(ii) Find the distance Quinn has walked when Martin catches up with him.

(iii) Find the distance between Quinn and Martin when Martin stops.

When answering this question, remember:
• You must change all times to seconds.
• You must use the equations for constant speed for Quinn's journey.

You must use both the equations for uniform acceleration and for constant speed for Martin's journey as shown below.

(a) To plot the graph we need to work out how long Martin takes to reach his maximum speed.
For the first part of Martin's journey:

$u = 0$ $v = u + at$
$a = 2.4$ $12 = 2.4t$
$v = 12$ $t = \dfrac{12}{2.4} = 5$ seconds

So Martin reaches 12 m/s after 5 seconds.
Because he doesn't start until 2 minutes after Quinn, ie 120 seconds, we plot (125, 12) on the graph.
We can now draw the graph, where T is the time Martin catches up with Quinn:

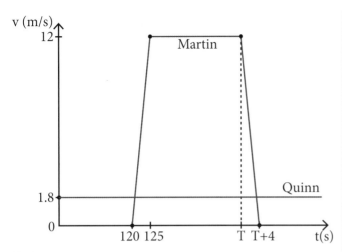

(b) (i) Let:
T = the time when Martin catches up with Quinn
D = the distance to this point
Now consider Martin and Quinn separately.

Quinn is travelling at constant speed, so:
$$\text{Distance} = \text{velocity} \times \text{time}$$
$$s = vt$$
So $$D = 1.8T$$

Martin: Consider each part of his journey separately.
For the first part:
u = 0 $v^2 = u^2 + 2as$
a = 2.4 $144 = 0 + 2 \times 2.4 \times s$
v = 12 $s = \dfrac{144}{4.8} = 30 \text{ m}$

In the second part, t = T – 125
$$\text{Distance} = \text{velocity} \times \text{time}$$
$$s = vt$$
So $$D = 12(T - 125)$$
Therefore the total distance travelled by Martin is:
$$D = 12(T - 125) + 30$$
$$D = 12T - 1500 + 30$$
$$D = 12T - 1470$$

However we know that the distance travelled by Martin is equal to the distance travelled by Quinn, since this is the point where they meet. Therefore:
$$12T - 1470 = 1.8T$$
$$12T - 1.8T = 1470$$
$$10.2T = 1470$$
$$T = \frac{1470}{10.2}$$
T = 144 s, to the nearest integer

(ii) To find the distance Quinn has walked at time T, substitute T into either equation for D:
$$D = 1.8T$$
$$D = 1.8 \times 144$$
$$D = 259.2 \text{ m}$$

(iii) Consider the last part of Martin's journey. You need to find the distance he travels.
u = 12
v = 0 (because he stops)
t = 4
$$s = \frac{1}{2} t (u + v)$$
$$s = \frac{1}{2}(4)(12 + 0)$$
$$s = 24 \text{ m}$$

However, Quinn is still walking, so we need to find out how far he walks in these 4 seconds:
$$\text{Distance} = \text{velocity} \times \text{time}$$
$$s = vt$$
So $$D = 1.8 \times 4$$
$$D = 7.2 \text{m}$$
So the distance between them is:
$$= 24 - 7.2$$
$$= 16.8 \text{ m}$$

Exercise 15C:

1. Lily leaves home 4 minutes before Cadence and walks at a constant speed of 1.2 m/s. Cadence cycles after her accelerating from rest at 0.8 m/s² for 10s until she reaches a maximum velocity which she maintains until she overtakes Lily. She then immediately decelerates uniformly to rest 20 s after overtaking Lily, who continues to walk at 1.2 m/s.
(a) Draw a velocity/time graph to show these journeys.
(b) (i) Find Cadence's maximum velocity.
    (ii) Find the distance from home when Cadence overtakes Lily.
    (iii) Find Cadence's deceleration.
    (iii) Find the distance between Cadence and Lily when Cadence stops.

2. Lennon leaves home walking at a constant velocity of 1.5 m/s. 3 minutes later Sandra leaves home. He cycles from rest at an acceleration of 2 m/s² for 6 s until he reaches a maximum velocity which she maintains until he overtakes Lennon.
(a) Draw a velocity/time graph to show these journeys.
(b) (i) Find the time Lennon has walked when Sandra catches up with him.
    (ii) Find the distance from home when Sandra catches up with Lennon.
    Sandra then immediately decelerates uniformly to rest at 1.2 m/s² after overtaking Lennon, who continues to walk at 1.2 m/s.
    (iii) Find the distance between Sandra and Lennon when Sandra stops.

CHAPTER 16: CONSTANT ACCELERATION

16.1 Equations of Motion

These equations of motion were derived in Chapter 15.
Where:
u = initial velocity in m/s
v = final velocity in m/s
a = acceleration in m/s²
t = time in s
s = distance in m

Then:

1. $v = u + at$
2. $v^2 = u^2 + 2as$
3. $s = ut + \dfrac{1}{2}at^2$
4. $s = \dfrac{1}{2}t(u + v)$

The rules for using these equations are:
• List what you know and what you want to find,
• Choose the appropriate formula.

For example:
1. A body moving at a velocity of 3 m/s accelerates uniformly at 2 m/s² for 5 s. Find
(i) its final velocity
(ii) the distance it covers.

(i) We know: $u = 3$
$a = 2$
$t = 5$
We want to find $v = ?$
So we use: $v = u + at$
$v = 3 + 2 \times 5$
$v = 3 + 10$
$v = 13$ m/s

(ii) We know $u = 3$
$a = 2$
$t = 5$
We want to find $s = ?$
So we use: $s = ut + \dfrac{1}{2}at^2$
$s = 3(5) + \dfrac{1}{2}(2)(5)^2$
$s = 15 + 25$
$s = 40$ m

2. A body reaches a velocity of 24 m/s after accelerating uniformly for 8 s, during which time it travels 112 m. Find
(i) its initial velocity
(ii) its acceleration.

(i) We know $v = 24$
$t = 8$
$s = 112$
We want to find $u = ?$
So we use: $s = \dfrac{1}{2}t(u + v)$
$112 = \dfrac{1}{2} \times 8(u + 24)$
$112 = 4(u + 24)$
$112 = 4u + 96$
$112 - 96 = 4u$
$16 = 4u$
$u = 4$ m/s

(ii) We know: $u = 4$
$v = 24$
$t = 8$
We want to find $a = ?$
So we use: $v = u + at$
$24 = 4 + 8a$
$20 = 8a$
$a = 2.5$ m/s²

Exercise 16A:

1. A body accelerates uniformly from 3 m/s to 9 m/s in 4 s. Find
(i) its acceleration (ii) the distance covered.

2. A body accelerates uniformly from 2 m/s to 6 m/s in 6 s. Find
(i) its acceleration (ii) the distance covered.

3. A body accelerates uniformly for 4 s at 1.5 m/s². Its initial velocity is 8 m/s. Find
(i) its final velocity (ii) the distance covered.

4. A body initially moving at 10 m/s accelerates for 8 s covering 100 m. Find
(i) its final velocity (ii) its acceleration.

5. A body accelerates uniformly from 3 m/s to 12 m/s in covering 30 m. Find
(i) the time taken (ii) its acceleration.

6. A body accelerates uniformly from 6 m/s to 10 m/s at an acceleration of 0.4 m/s². Find
(i) the distance covered (ii) the time taken.

7. A body reaches a velocity of 4 m/s after accelerating for 12 s and covering 36 m. Find
(i) its initial velocity (ii) its acceleration.

8. A body accelerates uniformly from 4 m/s for 6 s at 3 m/s^2. Find
(i) its final velocity (ii) the distance travelled.

9. A body accelerates at 4 m/s^2 covering 152 m in reaching a velocity of 35 m/s. Find
(i) its initial velocity (ii) the time taken.

10. A body accelerates uniformly from 8 m/s for 4 s in covering 80 m. Find
(i) its acceleration (ii) its final velocity.

11. A body accelerates uniformly from 2 m/s at 3.5 m/s^2 in travelling 159.75 m. Find
(i) its final velocity (ii) the time taken.

12. A body accelerates at 1.5 m/s^2 for 2s acquiring a velocity of 10 m/s. Find
(i) its initial velocity (ii) the distance travelled.

13. A body accelerates uniformly from 9 m/s to 30 m/s in travelling 136.5 m. Find
(i) its acceleration (ii) the time taken.

14. A body accelerates uniformly from 9.9 m/s to 33 m/s in 6.5 s. Find
(i) its acceleration (ii) the distance travelled.

16.2 Deceleration

When a body slows down this is called a **deceleration**. A deceleration can be written as a negative acceleration. For example, if the deceleration is 2 m/s^2 then you would write a = –2 m/s^2.

The rules for solving problems involving deceleration are the same as before, except that you need to be careful to write the number correctly as positive or negative:
• List what you know and what you want to find.
• Write the deceleration, if given, as a **negative** acceleration.
• Choose the appropriate formula.
• Write the deceleration, if found, as a **positive** quantity.
For example:

A body travelling at 4 m/s comes to rest uniformly in 1.6 s. Find
(i) its deceleration
(ii) the distance travelled in coming to rest.

(i) The phrase 'coming to rest' means that the final velocity is 0 m/s. Therefore:
We know   u = 4
            v = 0
            t = 1.6

We want to find a = ?
So we use   v = u + at
                0 = 4 + 1.6a
            –1.6a = 4

$$a = \frac{4}{-1.6}$$

                a = –2.5

Remember the question asks us to find the **deceleration**. So the deceleration = 2.5 m/s^2

(ii) We know   u = 4
                 v = 0
                 t = 1.6

We want to find s = ?
So we use $s = \frac{1}{2} t(u + v)$

                $s = \frac{1}{2} \times 1.6(4 + 0)$

                s = 0.8 × 4
                s = 3.2 m

1. A body travelling at 12 m/s decelerates uniformly at 3 m/s^2 in 2 s. Find
(i) the final velocity (ii) the distance travelled.

2. A body decelerates uniformly at 2.5 m/s^2 to a velocity of 7 m/s in moving 48 m. Find
(i) the initial velocity (ii) the time taken.

3. A body travelling at 12 m/s comes to rest uniformly in moving 24 m. Find
(i) its deceleration (ii) the time taken.

4. A body travelling at 13 m/s slows down uniformly to 4 m/s at a deceleration of 6 m/s^2. Find
(i) the time taken
(ii) the distance travelled in coming to rest.

5. A body travelling at 3 m/s comes to rest uniformly in 2 s. Find
(i) its deceleration
(ii) the distance travelled in coming to rest.

6. A body travelling at 14 m/s decelerates uniformly at 2 m/s^2 in moving 33 m. Find
(i) the final velocity (ii) the the time taken.

7. A body decelerates uniformly at 4 m/s^2 to a velocity of 2 m/s in 1.5 s. Find
(i) the initial velocity (ii) the distance travelled.

8. A body travels 24 m in coming to rest uniformly in 2 s.
Find
(i) the initial velocity (ii) its deceleration.

16.3 Vertical Motion

All bodies fall vertically at the same acceleration due to gravity. This acceleration is called 'g'. In this study, **g is equal to 10 m/s².**
• **Bodies falling down** will **accelerate downwards** and so you take a = 10 m/s².
• **Bodies going up** will **slow down** and so you take a = –10 m/s².
For example:

1. A ball drops 1.4 m to the ground. Find
(i) the velocity with which it hits the ground
(ii) the time taken
(iii) the distance travelled in the 8th second.

(i) The word 'drops' means that the initial velocity is 0. The ball is falling down and so a = 10

We know $\quad\quad u = 0$
$\quad\quad\quad\quad\quad a = 10$
$\quad\quad\quad\quad\quad s = 1.4$

We want to find v = ?
So we use $\quad\quad v^2 = u^2 + 2as$
$\quad\quad\quad\quad\quad v^2 = 0 + 2(10)(1.4)$
$\quad\quad\quad\quad\quad v^2 = 28$
$\quad\quad\quad\quad\quad v = \sqrt{28}$
$\quad\quad\quad\quad\quad v = 5.29 \text{ m/s}$
(ii) We know $\quad u = 0$
$\quad\quad\quad\quad\quad a = 10$
$\quad\quad\quad\quad\quad s = 1.4$

We want to find t = ?

So we use $\quad\quad s = ut + \frac{1}{2}at^2$
$\quad\quad\quad\quad 1.4 = 0 + \frac{1}{2}(10)t^2$
$\quad\quad\quad\quad 1.4 = 5t^2$
$\quad\quad\quad\quad t^2 = \frac{1.4}{5}$
$\quad\quad\quad\quad t^2 = 0.28$
$\quad\quad\quad\quad t = \sqrt{0.28}$
$\quad\quad\quad\quad t = 0.529s$

(iii) To find the distance travelled in the 8th second you:
• find the distance travelled after 7s
• find the distance travelled after 8s
• subtract the answers

First we set $\quad\quad t = 7$
So we know $\quad\quad u = 0$
$\quad\quad\quad\quad\quad a = 10$
$\quad\quad\quad\quad\quad t = 7$

We want to find s = ?
So we use $\quad\quad s = ut + \frac{1}{2}at^2$
$\quad\quad\quad\quad s = 0 + \frac{1}{2}(10)(7)^2$
$\quad\quad\quad\quad s = 5 \times 49 = 245 \text{ m}$
Second, we set $\quad t = 8$
So we know $\quad\quad u = 0$
$\quad\quad\quad\quad\quad a = 10$
$\quad\quad\quad\quad\quad t = 8$
We want to find s = ?
So we use $\quad\quad s = ut + \frac{1}{2}at^2$
$\quad\quad\quad\quad s = 0 + \frac{1}{2}(10)(8)^2$
$\quad\quad\quad\quad s = 5 \times 64 = 320 \text{ m}$
Finally, we subtract
Distance travelled = 320 – 245 = 75 m

. . . .

2. A stone is thrown up from the ground with a velocity of 3.5 m/s. Find
(i) how high will it reach
(ii) the time it takes to return to the ground
(iii) between which times the stone is more than 0.4 m above the ground

(i) The stone will go up until it stops, ie when v =0
We know $\quad\quad u = 3.5$
$\quad\quad\quad\quad\quad v = 0$
$\quad\quad\quad\quad\quad a = -10$
We want to find s = ?
So we use $\quad\quad v^2 = u^2 + 2as$
$\quad\quad\quad\quad\quad v^2 = u^2 + 2as$
$\quad\quad\quad\quad 0 = 3.5^2 - 2(10)s$
$\quad\quad\quad 20s = 12.25$
$\quad\quad\quad\quad\quad s = \frac{12.25}{20}$
$\quad\quad\quad\quad\quad s = 0.6125 \text{ m}$

(ii) The time taken for the stone to reach its greatest height (from 3.5 m/s to 0 m/s) is the same as the time taken for it to return to the ground (from 0 m/s to 3.5 m/s). So you work out the time taken for the stone to reach its greatest height and then double this.
We know $\quad\quad u = 3.5$
$\quad\quad\quad\quad\quad v = 0$
$\quad\quad\quad\quad\quad a = -10$
We want to find t = ?
So we use $\quad\quad v = u + at$
$\quad\quad\quad\quad 0 = 3.5 - 10t$
$\quad\quad\quad 10t = 3.5$
$\quad\quad\quad\quad\quad t = 0.35 \text{ s to reach the greatest height}$
So the time for it to return to the ground is
$\quad\quad\quad 0.35 \times 2 = 0.7s$

(iii) We need to find the times when s = 0.4

We know

$u = 3.5$
$a = -10$
$s = 0.4$

We want to find t = ?

So we use

$$s = ut + \frac{1}{2}at^2$$

$$0.4 = 3.5t - \frac{1}{2} \times 10\, t^2$$

$$0.4 = 3.5t - 5t^2$$

This is a quadratic equation. You must bring all the terms to one side and then solve it using the quadratic formula

So we have:

$$5t^2 - 3.5t + 0.4 = 0$$

Thus

$a = 5$
$b = -3.5$
$c = 0.4$

Using

$$t = \frac{-b \pm \sqrt{b^2 - 4ac}}{2a}$$

$$t = \frac{3.5 \pm \sqrt{3.5^2 - 4 \times 5 \times 0.4}}{2 \times 5}$$

$$t = \frac{3.5 \pm \sqrt{4.25}}{10}$$

$$t = \frac{3.5 + \sqrt{4.25}}{10} \text{ or } \frac{3.5 - \sqrt{4.25}}{10}$$

$$t = \frac{5.562}{10} \text{ or } \frac{1.428}{10}$$

$$t = 0.556\text{ s or } 0.143\text{ s}$$

So the answer is that the the stone is more than 0.4 m above the ground between 0.143 s and 0.556s .

Exercise 16C:

1. A ball takes 2.5s to drop to the floor. Find
   (i) the velocity with which it hits the floor
   (ii) how far it falls.

2. A stone is thrown up from the floor at 1.4 m/s. Find
   (i) how high it rises
   (ii) the time it takes to reach this height.

3. A ball drops 1.84 m to the floor. Find
   (i) the velocity with which it hits the floor
   (ii) the time it takes to reach the floor.

Exercise 16C...

4. A stone is thrown up from the floor at 2.3 m/s. Find
   (i) how high it rises
   (ii) the time it takes to return to the floor.

5. A ball is thrown down with a velocity of 1.3 m/s. Find
   (i) the distance travelled in the third second
   (ii) the distance travelled after 5 seconds
   (iii) its velocity after 5 seconds.

6. A stone is thrown up from the floor at 3.8 m/s. Find
   (i) its greatest height
   (ii) the times when it is 0.5 m above the floor.

7. A ball is thrown with a velocity of 0.4 m/s. Find
   (i) the time taken to fall 3 m
   (ii) the further distance travelled in the next 2 seconds.

8. A stone is thrown up from the ground at 1.56 m/s. Find
   (i) its speed and direction after 0.2s
   (ii) the time to reach its greatest height
   (iii) its greatest height
   (iv) the times between which the stone is more than 8 cm above the ground.

16.4 Harder Questions

The same methods can be use to solve more difficult problems. For example:

1. A stone is dropped from a shelf 2 m above the ground At the same moment another stone is thrown up from the ground at 6.2 m/s. Find
(i) when they pass each other
(ii) how far above the ground they are when they pass
(iii) the distance between them after 0.2 s.

(i) Consider each stone separately.
Let   t = the time when they pass
and   h = the height above the ground
The stone that drops will have fallen a distance of
$$s = 2 - h$$

Firstly, consider the stone dropping
We know

$u = 0$
$a = 10$ (because it is falling down)
$s = 2 - h$

We want to find t = ? (ie, the time they pass)

We use $\quad s = ut + \dfrac{1}{2}at^2$

$$2 - h = 0 + \dfrac{1}{2}10t^2$$

$$2 - h = 5t^2$$

Next, consider the stone thrown up

We know $\quad u = 6.2$

$\quad a = -10$ (because it is going up)

$\quad s = h$

We want to find t = ? (ie, the time they pass)

We use $s = ut + \dfrac{1}{2}at^2$

$$h = 6.2t - \dfrac{1}{2}10t^2$$

$$h = 6.2t - 5t^2$$

$$5t^2 = 6.2t - h$$

You can put the two expressions for $5t^2$ equal to get:

$$6.2t - h = 2 - h$$

which gives $\quad 6.2t = 2$

$$\text{So } t = \dfrac{2}{6.2}$$

$$t = 0.323s$$

(ii) Since we now know t, we can substitute t = 0.323 into either equation:

$$2 - h = 5t^2$$

Rearrange: $2 - 5t^2 = h$

$$h = 2 - 5t^2$$

$$h = 2 - 5(0.323)^2$$

$$h = 1.48 \text{ m}$$

(iii) Substitute t = 0.2 into each equation to find the height of each stone at this time, and then subtract the answers to find the distance between them:

Dropping stone: $h = 2 - 5t^2$

$$h = 2 - 5(0.2)^2$$

$$h = 1.8 \text{ m}$$

Stone going up: $h = 6.2t - 5t^2$

$$h = 6.2(0.2) - 5(0.2)^2$$

$$h = 1.04 \text{ m}$$

So the distance between them

$$s = 1.8 - 1.04$$

$$= 0.76 \text{ m}$$

....

2. A ball is thrown up at 6 m/s from the top of a cliff 4.6 m above the ground. Find
(i) the greatest height above the ground that it reaches
(ii) the total time it takes it to reach the ground
(iii) the times between which it is 5.8 m above the ground

(i) We know $\quad u = 6$

$\quad v = 0$

$\quad a = -10$

We want to find s = ?

So we use $\quad v^2 = u^2 + 2as$

$$0 = 6^2 - 2(10)s$$

$$20s = 36$$

$$s = \dfrac{36}{20}$$

$$s = 1.8 \text{ m}$$

However, this is the height above the point it was thrown from. But it was thrown from the top of a cliff. So the greatest height above the ground is

$$= 1.8 + 4.6$$

$$= 6.4 \text{ m}$$

(ii) There are two time intervals - the time interval when the ball is moving upwards, and the time interval when it is moving downwards. You need to work out the times for both and add up the answers.

Firstly, from the top of the cliff to the greatest height:

We know $\quad u = 6$

$\quad v = 0$

$\quad a = -10$

We want to find t = ?

So we use $\quad v = u + at$

$$0 = 6 - 10t$$

$$10t = 6$$

So $\quad t = 0.6s$ to reach the greatest height

Then, from the greatest height to the ground:

We know $\quad u = 0$

$\quad a = 10$

$\quad s = 6.4$ (the greatest height)

We want to find t = ?

We use $\quad s = ut + \dfrac{1}{2}at^2$

$$6.4 = (0 \times t) + \left(\dfrac{1}{2} \times 10t^2\right)$$

$$6.4 = \dfrac{1}{2} \times 10t^2$$

$$6.4 = 5t^2$$

$$1.28 = t^2$$

$$1.131 = t$$

Add the two time intervals together to get:

$$\text{Total time} = 0.6 + 1.131$$

$$= 1.731 \text{ s}$$

77

(iii) The ball is thrown from the top of a cliff, so you must first subtract 4.6 from 5.8 to get s = 1.2

We know $\quad$ u = 6

$\quad$ a = −10

$\quad$ s = 1.2

We want to find $\;$ t = ?

We use $\qquad s = ut + \dfrac{1}{2}at^2$

$$1.2 = 6t + \dfrac{1}{2} \times -10t^2$$

$$1.2 = 6t - 5t^2$$

This is a quadratic equation. You must bring all the terms to one side and then solve it using the quadratic formula.

So we have:

$$5t^2 - 6t + 1.2 = 0$$

Thus $\qquad$ a = 5

$\qquad$ b = −6

$\qquad$ c = 1.2

Using $\qquad t = \dfrac{-b \pm \sqrt{b^2 - 4ac}}{2a}$

$$t = \dfrac{6 \pm \sqrt{6^2 - 4 \times 5 \times 1.2}}{2 \times 5}$$

$$t = \dfrac{6 \pm \sqrt{12}}{10}$$

$$t = \dfrac{6 + \sqrt{12}}{10} \text{ or } \dfrac{6 - \sqrt{12}}{10}$$

$$t = \dfrac{9.464}{10} \text{ or } \dfrac{2.536}{10}$$

Thus $\qquad$ t = 0.946 s or 0.254 s

So the ball is 5.8 m above the ground at times t = 0.254 s and t = 0.946 s.

Exercise 16D:

1. A stone is dropped from a shelf 1.4 m above the ground. At the same moment another stone is thrown up from the ground at 5.3 m/s.
Find
(i) when they pass each other.
(ii) how far above the ground they are when they pass.
(iii) the distance between them after 0.15 s.

2. A stone falls from a cliff h metres high to the ground.
0.2 s later another stone is thrown down from the cliff at 2.3 m/s. They hit the ground at the same time. Find
(i) when they hit the ground.
(ii) the height, h, of the cliff.
(iii) the velocity of each stone when they hit the ground.

3. A ball is thrown up at 5 m/s from the top of a cliff 2.8 m above the ground. Find
(i) the greatest height above the ground that it reaches
(ii) the total time it takes it to reach the ground.
(iii) the velocity with which it hits the ground.
(iv) the times between which it is 3.5 m above the ground.

4. A ball is thrown up at 3.2 m/s from the top of a cliff 8.4 m above the ground. Find
(i) the greatest height above the ground that it reaches
(ii) the total time it takes it to reach the ground.
(iii) the velocity with which it hits the ground.
(iv) the times between which it is 8.6 m above the ground.

5. A ball falls from a cliff 2.3 m high to the ground 0.2 s later another stone is thrown up from the ground at 4.5 m/s. Find
(i) when they pass.
(ii) the height above the ground when they pass.
(iii) how many seconds between the ball hitting the ground and the stone hitting the ground.

6. A stone is dropped from a cliff 5 m above the ground
At the same moment another stone is thrown up from the ground at 8 m/s. Find
(i) when they pass each other.
(ii) how far above the ground they are when they pass.
(iii) the distance between them after 0.4s.

CHAPTER 17: NEWTON'S LAWS

17.1 Definitions

Newton's three laws of motion state:

Law 1: Every body remains stationary or in constant or uniform motion unless acted upon by an external force.

Law 2: The force is proportional to the acceleration. The force, F, is worked out using the formula:

$$F = ma$$

where   m   is the mass of the object
           a   is the acceleration

The units of force are Newtons (N) when the mass is in kg and the acceleration is in m/s^2.

Law 3: To every action there is an equal and opposite reaction.

17.2 Acceleration Forces

You can use Newton's Laws to solve problems involving forces and accelerating objects. For example:

A force 34N causes a mass of M kg to accelerate at 1.6 m/s^2. Find the value of M.

$$F = ma$$
$$34 = M \times 1.6$$

So $M = \dfrac{34}{1.6}$

            $M = 21.25$ kg

Exercise 17A:

1. Find the force which will cause a mass of 8 kg to accelerate at 2.5 m/s^2.

2. A force 24N causes a mass of M kg to accelerate at 4 m/s^2. Find the value of M.

3. A force of 32N is used to accelerate a mass of 6.4 kg. Find its acceleration.

4. Find the force which will cause a mass of 20 kg to accelerate at 1.2 m/s^2.

5. A force 28.8N causes a mass of M kg to accelerate at 3.6 m/s^2. Find the value of M.

6. A force of 21N is used to accelerate a mass of 14 kg. Find its acceleration.

Exercise 17A...

7. Find the force which will cause a mass of 8 kg to accelerate at 2.8 m/s^2.

8. A force 18N causes a mass of M kg to accelerate at 0.6 m/s^2. Find the value of M.

Sometimes you have to work out the information you need from other information you are given. For example:

(i) Find the force needed to accelerate a mass of 3kg from 3 m/s to 8 m/s in 4s.
(ii) Find the distance travelled.

(i) To solve this question, we must first find the acceleration, and then use F=ma.

We know   u = 3
               t = 4
               v = 8

We want to find   a = ?

So we use   v = u + at
               8 = 3 + 4a
             4a = 5
              a = 1.25 m/s^2

Then   F = ma
              F = 3 × 1.25
                = 3.75 N

(ii) We know   u = 3
               t = 4
               v = 8

We want to find   s = ?

So we use $s = \dfrac{1}{2} t(u + v)$

              $s = \dfrac{1}{2} \times 4(3 + 8)$

             s = 2 × 11
             s = 22 m

Exercise 17B:

1. (i) Find the force needed to accelerate a mass of 4 kg from 4 m/s to 13 m/s in 6 s.
   (ii) Find the distance travelled.

2. (i) Find the force needed to accelerate a mass of 9 kg from 8 m/s to 14 m/s in travelling 22 m.
   (ii) Find the time taken.

Exercise 17B...

3. (i) Find the force needed to accelerate a mass of 6 kg to a velocity of 43 m/s in 8 s in travelling 200m.
(ii) Find the initial velocity.

4. (i) Find the force needed to accelerate a mass of 8 kg initially moving at a velocity of 2 m/s in 5s in travelling 97.5 m.
(ii) Find the final velocity.

5. (i) Find the force needed to accelerate a mass of 12 kg from 13 m/s to 25 m/s in travelling 114 m.
(ii) Find the time taken.

6. (i) Find the force needed to accelerate a mass of 16 kg from 4 m/s to 11 m/s in 5 s.
(ii) Find the distance travelled.

7. (i) Find the force needed to accelerate a mass of 6 kg initially moving at a velocity of 3 m/s in 9s in travelling 189 m.
(ii) Find the final velocity.

8. (i) Find the force needed to accelerate a mass of 15 kg to a velocity of 18 m/s in 4s in travelling 52 m.
(ii) Find the initial velocity.

17.3 Deceleration Force

You can use the same method to solve problems involving decelerating objects.For example:
(i) Find the force needed to bring a body of mass 7 kg moving at 14 m/s to rest in 3.5 s.
(ii) Find the distance covered.

(i) To solve this question, we must first find the acceleration, and then use F=ma.
We know $\quad u = 14$
$\qquad\qquad v = 0$ (as it comes to rest)
$\qquad\qquad t = 3.5$
We want to find $a = ?$
So we use $\qquad v = u + at$
$\qquad\qquad 0 = 14 + 3.5a$
$\qquad -3.5a = 14$
$\qquad\qquad a = -4 \text{ m/s}^2$
Then $\qquad\quad F = ma$
$\qquad\qquad F = 7 \times -4 = -28 \text{ N}$

Because the question asked for the force, you must give the positive value as the answer. Therefore the answer is that the decelerating force is 28 N.

(ii) We know $\qquad u = 14$
$\qquad\qquad\quad t = 3.5$
$\qquad\qquad\quad v = 0$
We want to find $s = ?$
So we use $\qquad s = \dfrac{1}{2} t(u + v)$
$\qquad\qquad\quad s = \dfrac{1}{2} \times 3.5(14 + 0)$
$\qquad\qquad\quad s = 3.5 \times 7$
$\qquad\qquad\quad s = 24.5 \text{ m}$

Exercise 17C:

1. (i) Find the force needed to bring a body of mass 7 kg to rest in 2 s and in 4 m.
(ii) Find the initial velocity.

2. (i) Find the force needed to bring a body of mass 4 kg moving at 6 m/s to rest in 4.5 m.
(ii) Find the time taken.

3. (i) Find the force needed to bring a body of mass 9 kg moving at 12 m/s to rest in 3 s.
(ii) Find the distance covered.

4. (i) Find the force needed to bring a body of mass 6 kg to rest in 2 s and in 7 m.
(ii) Find the initial velocity.

5. (i) Find the force needed to bring a body of mass 8 kg moving at 5 m/s to rest in 10 m.
(ii) Find the time taken.

6. (i) Find the force needed to bring a body of mass 7 kg moving at 3 m/s to rest in 3 m.
(ii) Find the time taken.

7. (i) Find the force needed to bring a body of mass 4 kg moving at 9 m/s to rest in 3 s.
(ii) Find the distance covered.

8. (i) Find the force needed to bring a body of mass 8 kg to rest in 2 s and in 5 m.
(ii) Find the initial velocity.

9. (i) Find the force needed to bring a body of mass 6 kg moving at 6 m/s to rest in 12 m.
(ii) Find the time taken.

10. (i) Find the force needed to bring a body of mass 9 kg moving at 7 m/s to rest in 5 s.
(ii) Find the distance covered.

CHAPTER 18: FORCES

18.1 Resolving Forces

A force is a vector with both magnitude and direction. You can split a force into its two perpendicular components by using trigonometry. This process is known as 'resolving the forces'. For example:

1. Work out the horizontal and vertical component of the force shown below.

You can resolve the 4 N force into its two components (the horizontal component R_H and the vertical component R_V) as shown below.

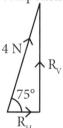

You can now work out the values of R_H and R_V by using trigonometry:

So $\quad \sin 75 = \dfrac{R_V}{4}$

$\quad\quad R_V = 4 \sin 75$
$\quad\quad R_V = 3.86 \text{ N}$

And $\quad \cos 75 = \dfrac{R_H}{4}$

$\quad\quad R_H = 4 \cos 75$
$\quad\quad R_H = 1.04 \text{ N}$

....

2. Work out the horizontal and vertical component of the force shown below.

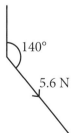

You can resolve the 4 N force into its two components (the horizontal component R_H and the vertical

component R_V) as shown below. You must work with the acute angle of 40° as shown.

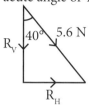

You can now work out the values of R_H and R_V by using trigonometry :

So $\quad \sin 40 = \dfrac{R_H}{5.6}$

$\quad\quad R_H = 5.6 \sin 40$
$\quad\quad R_H = 3.60 \text{ N}$

And $\quad \cos 40 = \dfrac{R_V}{5.6}$

$\quad\quad R_V = 5.6 \cos 40$
$\quad\quad R_V = 4.29 \text{ N}$

Exercise 18A: *Work out the horizontal and vertical components of the forces shown below.*

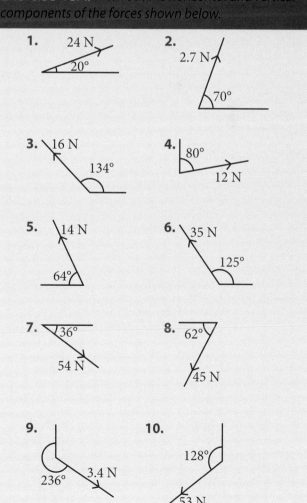

18.2 Resultant Of Parallel Forces

The **resultant** of a set of forces is the **single** force that is **equivalent** to all the **other** forces. You must give both its magnitude and also its direction since force is a **vector**.

To find the resultant of **parallel forces** you:
• Add them if they act in the same direction,
• Subtract them if they act in opposite directions.
For example:

1. Find the resultant of the following forces:

7 N
6 N
4 N

The resultant force is 7 + 6 + 4 = 17 N acting horizontally to the right.

....

2. Find the resultant of the following forces:

4 N
10 N
3 N

The resultant force is 10 − 4 − 3 = 3 N acting horizontally to the left.

18.3 Resultant Of Non-Parallel Forces Acting At A Point

When you have two or more non-parallel forces acting at a point you can find the resultant of these forces by:

• Resolving each force if necessary.

• Finding the resultant force in each of two perpendicular directions. When the forces are acting on a **horizontal plane**, you resolve the forces **horizontally** and **vertically**. When the forces are acting on an **inclined plane**, you resolve the forces **along** the plane and **perpendicular** to the plane.

• Finding the magnitude of the resultant force by using Pythagoras' theorem.

• Finding the direction of the resultant force by using trigonometry.

For example:

1. Find the magnitude and direction of the resultant of the forces shown below:

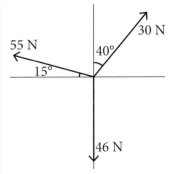

You must first resolve the 55 N and the 30 N forces into their horizontal and vertical components. You do not need to resolve the 46 N force as it is a vertical force. This gives the following:

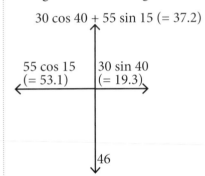

$30 \cos 40 + 55 \sin 15 \ (= 37.2)$

$55 \cos 15$ $30 \sin 40$
$(= 53.1)$ $(= 19.3)$

46

The resultant horizontal force is:
$53.1 − 19.3 = 33.8$ acting to the left
The resultant vertical force is:
$46 − 37.2 = 8.8$ acting down
You can then calculate the magnitude and direction of this resultant force as follows:

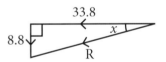

33.8
8.8
x
R

First find R: $R^2 = 33.8^2 + 8.8^2$
$R^2 = 1219.88$
$R = \sqrt{1219.88}$
$R = 34.9$ N, to 3 significant figures

Then find x: $\tan x = \dfrac{8.8}{33.8}$
$\tan x = 0.2603...$
$x = 14.6°$, to 3 significant figures

Answer: A force of 34.9 N acting at 14.6° to the horizontal, as shown above.

Exercise 18B: *Find the magnitude and direction of the resultant of the forces shown below.*

1.

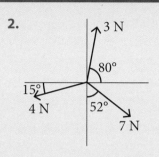

2.

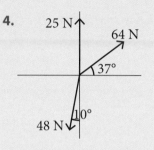

3.

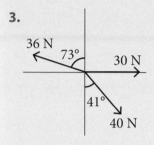

4.

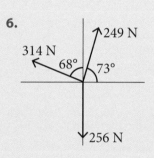

5.

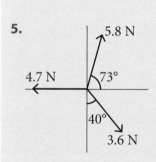

6.

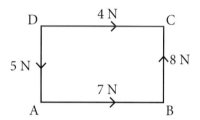

7.

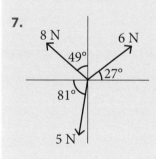

8.

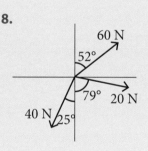

9.

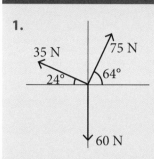

10.

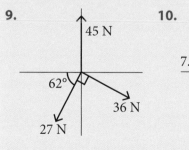

18.4 Resultant Of Forces Acting Along Different Sides Of A Rectangle

The rules are similar:
- Find the resultant force horizontally and vertically.
- Find the magnitude of the resultant force by using Pythagoras' theorem.
- Find the direction of the resultant force by using trigonometry.

For example:

Find the magnitude and direction of the resultant of the forces shown below:

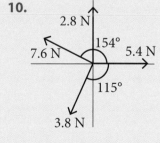

The resultant horizontal force is:

$$7 + 4 = 11 \text{ acting to the right}$$

The resultant vertical force is:

$$8 - 5 = 3 \text{ acting up}$$

You can then calculate the magnitude and direction of this resultant force as follows:

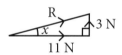

First find R: $R^2 = 3^2 + 11^2$

$R^2 = 130$

$R = \sqrt{130}$

$R = 11.4$ N, to 3 significant figures

Then find x: $\tan x = \dfrac{3}{11}$

$\tan x = 0.2727...$

$x = 15.3°$, to 3 significant figures

Answer: A force of 11.4 N acting at 15.3° to the horizontal, as shown above.

Exercise 18C:

Use the rectangle below to find the magnitude and direction of the resultant of the forces in each question below.

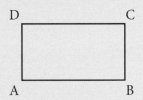

1. 7 N acting along DC and
   9 N acting along CB and
   4 N acting along AB and
   2 N acting along AD

2. 8 N acting along DC and
   5 N acting along BC and
   3 N acting along BA and
   6 N acting along AD

3. 16 N acting along DC and
   15 N acting along BC and
   14 N acting along BA and
   12 N acting along DA

4. 25 N acting along DC and
   10 N acting along BC and
   30 N acting along AB and
   14 N acting along DA

5. 7 N acting along DC and
   4 N acting along CB and
   9 N acting along BA and
   5 N acting along DA

6. 6.8 N acting along CD and
   2.7 N acting along CB and
   3.2 N acting along AB and
   5 N acting along AD

7. 24 N acting along DC and
   5 N acting along BC and
   13 N acting along BA and
   19 N acting along AD

8. 16 N acting along CD and
   5.8 N acting along CB and
   9.2 N acting along BA and
   7.4 N acting along DA

Exercise 18C...

9. 25 N acting along DC and
   22 N acting along CB and
   14 N acting along BA and
   16 N acting along AD

10. 27 N acting along DC and
    23 N acting along BC and
    14 N acting along BA and
    35 N acting along DA

18.5 Resultant Of Forces Acting Along An Inclined Plane

The rules in this case are:
- Find the resultant force along the plane and perpendicular to the plane.
- Find the magnitude of the resultant force by using Pythagoras' theorem.
- Find the direction of the resultant force by using trigonometry.

For example:

Find the magnitude and direction of the resultant of the forces shown below.

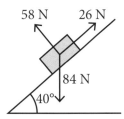

You do not need to resolve either the 58 N or the 26 N since they are already perpendicular or parallel to the plane. You resolve the 84 N force as follows:

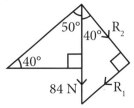

This gives:

So $\qquad \sin 40 = \dfrac{R_1}{84}$

$\qquad\qquad R_1 = 84 \sin 40$
$\qquad\qquad R_1 = 54$ N

And $\qquad \cos 40 = \dfrac{R_2}{84}$

$\qquad\qquad R_2 = 84 \cos 40$
$\qquad\qquad R_2 = 64.3$ N

You then get:

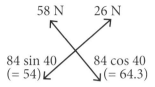

58 N   26 N

84 sin 40   84 cos 40
(= 54)   (= 64.3)

The resultant force along the plane is:

    54 − 26 = 28 acting down the plane

The resultant force perpendicular to the plane is

    64.3 − 58 = 6.3 acting down

You can then calculate the magnitude and direction of this resultant force as follows:

28 N x

R

6.3 N

First find R: $R^2 = 6.3^2 + 28^2$

            $R^2 = 823.69$

            $R = \sqrt{823.69}$

            $R = 28.7$ N

Then find x: $\tan x = \dfrac{6.3}{28}$

           $\tan x = 0.225\ldots$

              $x = 12.7°$, to 3 significant figures

Answer: A force of 28.7 N acting at 12.7° to the plane, as shown above.

Exercise 18D: *Find the magnitude and direction of the resultant of the forces shown below:*

1.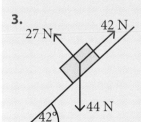
42 N   53 N

35°   60 N

2.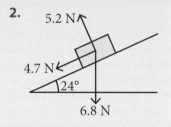
5.2 N

4.7 N

24°

6.8 N

3.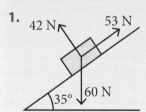
27 N   42 N

44 N

42°

4.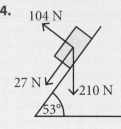
104 N

27 N   210 N

53°

5.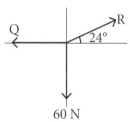
3.6 N   4.8 N

28°

5.2 N

6.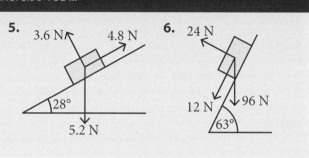
24 N

12 N   96 N

63°

18.6 Equilibrium Of Forces

When a set of forces is in **equilibrium** then the resultant force in each of two perpendicular directions will be **0**. You can use this fact to find unknown forces. The rules for finding unknown forces when all the forces are in equilibrium are:

- Resolve each force if necessary.
- Equate the forces in each of the two perpendicular directions.

For example:

Find the unknown forces below, given that the forces are in equilibrium:

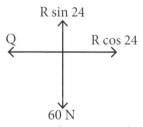

Q   R

24°

60 N

You do not need to resolve either the 60 N or the Q N forces as they are already in perpendicular directions. You resolve the RN force to get:

R sin 24

Q   R cos 24

60 N

You can then equate the vertical forces to get:

     R sin 24 = 60

         $R = \dfrac{60}{\sin 24}$

         R = 147.5 N

You can then equate the horizontal forces to get:

         Q = R cos 24

         Q = 147.5 cos 24

         Q = 134.7 N

Exercise 18E: *Find the unknown forces in the questions below, given that the forces are in equilibrium.*

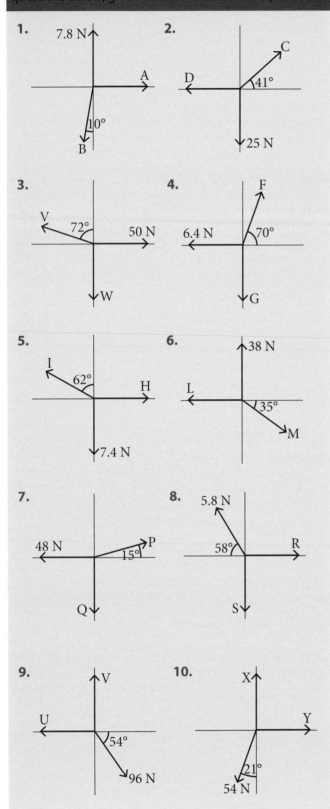

1. 7.8 N, A, 10°, B

2. C, D, 41°, 25 N

3. V, 72°, 50 N, W

4. F, 6.4 N, 70°, G

5. I, 62°, H, 7.4 N

6. 38 N, L, 35°, M

7. 48 N, P, 15°, Q

8. 5.8 N, 58°, R, S

9. V, U, 54°, 96 N

10. X, Y, 21°, 54 N

18.7 Forces On An Inclined Plane

You can also find unknown forces on an inclined plane, if the forces are in equilibrium. For example:

Find the unknown forces below, given that the forces are in equilibrium:

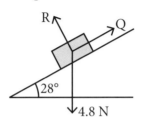

R, Q, 28°, 4.8 N

You should resolve the forces along the plane and perpendicular to the plane. Thus you do not need to resolve either R or Q.
You only need to resolve the 4.8 N force to get:

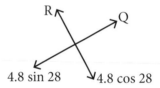

R, Q, 4.8 sin 28, 4.8 cos 28

You can then equate the forces to get:

$$R = 4.8 \cos 28$$
$$R = 4.24 \text{ N}$$

and
$$Q = 4.8 \sin 28$$
$$Q = 2.25 \text{ N}$$

Exercise 18F: *Find the unknown forces in the questions below, given that the forces are in equilibrium:*

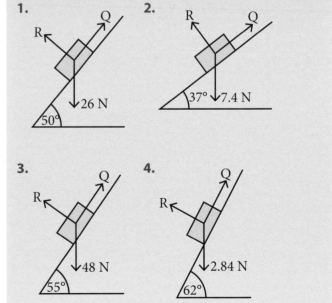

1. Q, R, 26 N, 50°

2. R, Q, 37°, 7.4 N

3. Q, R, 48 N, 55°

4. Q, R, 2.84 N, 62°

5.

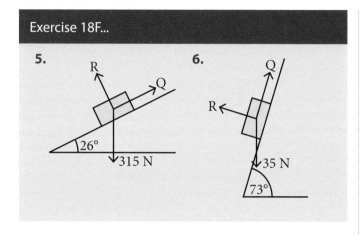

R

Q

26°

↓315 N

6.

Q

R←

↓35 N

73°

18.8 Harder Questions

You may be asked more difficult questions involving equilibrium. For example:

Find the unknown forces below, given that the forces are in equilibrium:

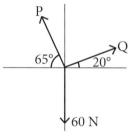

P

Q

65°

20°

↓60 N

You do not need to resolve the 60 N as it acts vertically. You resolve the two unknown forces force to get:

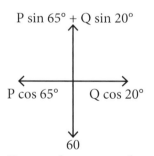

$P \sin 65° + Q \sin 20°$

$P \cos 65°$ | $Q \cos 20°$

60

You can then equate the vertical forces to get:

$$60 = P \sin 65° + Q \sin 20° \qquad [1]$$

And you can equate the horizontal forces to get:

$$P \cos 65° = Q \cos 20° \qquad [2]$$

You can rearrange [2] to get:

$$P = \frac{Q \cos 20°}{\cos 65°} \qquad [3]$$

Susbtituting [3] into [1] gives:

$$Q\left(\frac{\cos 20° \sin 65°}{\cos 65°}\right) + Q \sin 20° = 60$$

$$Q\left(\frac{\cos 20° \sin 65°}{\cos 65°} + \sin 20°\right) = 60$$

$$Q = 60 \div \left(\frac{\cos 20° \sin 65°}{\cos 65°} + \sin 20°\right)$$

$$Q = 25.45 \text{ N}$$

Substitute this value back into [3] to find P:

$$P = \frac{Q \cos 20°}{\cos 65°}$$

$$P = \frac{25.45 \cos 20°}{\cos 65°}$$

$$P = 56.6 \text{ N}$$

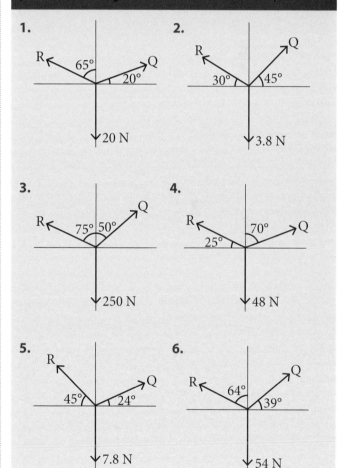

1.

R←

65°

20°

Q

↓20 N

2.

R

Q

30°

45°

↓3.8 N

3.

R←

75°

50°

Q

↓250 N

4.

R←

70°

25°

Q

↓48 N

5.

R

45°

24°

Q

↓7.8 N

6.

R←

64°

39°

Q

↓54 N

87

CHAPTER 19: VECTORS

19.1 Magnitude and Direction

Vectors are quantities with both **magnitude** and **direction**. For example:
   force, acceleration, velocity, displacement

Scalars are quantities with **only magnitude**. For example:
   mass, time, speed, distance

Vectors can be written in the form **ai + bj** where **i** and **j** denote unit vectors parallel to a set of standard x–y axes.

Look at the vector **ai + bj** below:

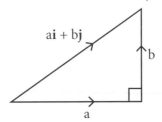

You can find the **magnitude of the vector** R, by using Pythagoras' theorem as follows:

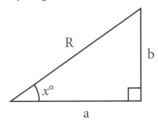

$$R^2 = a^2 + b^2$$
$$R = \sqrt{a^2 + b^2}$$

You can find the direction by using trigonometry as follows:

$$\tan x = \frac{b}{a}$$
$$x = \tan^{-1} \frac{b}{a}$$

For example:

Find the magnitude and direction of the force $-7\mathbf{i} - 4\mathbf{j}$

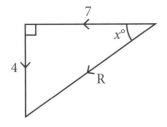

Magnitude: $R = \sqrt{(-7)^2 + (-4)^2}$
               $R = \sqrt{65}$
               $R = 8.06$ N

Direction: $\tan x = \frac{4}{7}$
               $x = \tan^{-1} \frac{4}{7}$
               $x = 29.7°$ with the horizontal

Exercise 19A: *Find the magnitude and direction of the following forces.*

1. $4\mathbf{i} + 5\mathbf{j}$	**2.** $7\mathbf{i} - 2\mathbf{j}$	**3.** $-3\mathbf{i} + 6\mathbf{j}$
4. $8\mathbf{i} + 15\mathbf{j}$	**5.** $-\mathbf{i} + 3\mathbf{j}$	**6.** $5\mathbf{i} + 2\mathbf{j}$
7. $12\mathbf{i} - 5\mathbf{j}$	**8.** $4\mathbf{i} + 2\mathbf{j}$	**9.** $5\mathbf{i} + 6\mathbf{j}$
10. $7\mathbf{i} - 3\mathbf{j}$		

19.2 Finding Speed And Distance

The **speed** of a body is the magnitude of the velocity. The **distance** a body travels is the magnitude of the displacement. For example:

1. A body moves with a velocity of $(6\mathbf{i} - 7\mathbf{j})$ m/s

Find its speed.
Magnitude: $R = \sqrt{6^2 + (-7)^2}$
               $R = \sqrt{85}$
               $R = 9.22$ m/s

2. A body has displacement $(-4\mathbf{i} + 2\mathbf{j})$ m from O. Find its distance from O.

Magnitude: $R = \sqrt{(-4)^2 + 2^2}$
               $R = \sqrt{20}$
               $R = 4.47$ m

Exercise 19B: *Find the magnitude and direction of the following forces.*

1. A body moves with a velocity of (6**i** – 8**j**) m/s. Find its speed.

2. A body moves with a velocity of (– 4**i** + 7**j**) m/s. Find its speed.

3. A body moves with a velocity of (– 2**i** – 8**j**) m/s. Find its speed.

4. A body has displacement (–**i** + 2**j**) m from O. Find its distance from O.

5. A body has displacement (24**i** – 10**j**) m from O. Find its distance from O.

6. A body has displacement (9**i** + 7**j**) m from O. Find its distance from O.

19.3 Resultant Of Forces

You can find the resultant of forces written as vectors by adding the **i** components together and adding the **j** components together. For example:

The 3 forces (4**i** – 6**j**) N, (– 2**i** + 5**j**) N and (– 3**i** + 4**j**) N act on a body. Find the resultant of these forces.

$$4\mathbf{i} - 6\mathbf{j}$$
$$-2\mathbf{i} + 5\mathbf{j}$$
$$+\ \underline{-3\mathbf{i} + 4\mathbf{j}}$$
$$-\mathbf{i} + 3\mathbf{j}$$

Answer: (–**i** + 3**j**) N

Exercise 19C:

1. The 3 forces (4**i** – 2**j**) N, (5**i** + 3**j**) N and (– 3**i** +**j**) N act on a body. Find the resultant of these forces.

2. The 3 forces (**i** + 3**j**) N, (–4**i** – 6**j**) N and (–2**i** + 5**j**) N act on a body. Find the resultant of these forces.

3. The 3 forces (3**i** – 2**j**) N, (–**i** – 3**j**) N and (–7**i** + 4**j**) N act on a body. Find the resultant of these forces.

4. The 3 forces (**i** – 2**j**) N, (–3**i** + 4**j**) N and (2**i** – **j**) N act on a body. Find the resultant of these forces.

5. The 3 forces (2**i** + 5**j**) N, (4**i** – 6**j**) N and (–3**i** + **j**) N act on a body. Find the resultant of these forces.

6. The 3 forces (**i** – 3**j**) N, (–2**i** + 4**j**) N and (5**i** – 2**j**) N act on a body. Find the resultant of these forces.

7. The 3 forces (–3**i** – 5**j**) N, (2**i** – 3**j**) N and (–**i** + 4**j**) N act on a body. Find the resultant of these forces.

Exercise 19C...

8. The 3 forces (–**i** – 4**j**) N, (3**i** – **j**) N and (–5**i** + 3**j**) N act on a body. Find the resultant of these forces.

9. The 3 forces (4**i** + 2**j**) N, (–**i** + 6**j**) N and (3**i** – 5**j**) N act on a body. Find the resultant of these forces.

10. The 3 forces (–3**i** + 4**j**) N, (5**i** + 2**j**) N and (–**i** + 6**j**) N act on a body. Find the resultant of these forces.

19.4 Finding An Unknown Force When You Are Given The Resultant Force

Sometimes you know the resultant force, and need to find an unknown force. The rules are:
• Add the forces you are given.
• Subtract this answer from the resultant force.

For example:

The forces **P** N, (–**i** + 6**j**) N and (3**i** – 5**j**) N act on a body. The resultant of these forces is (7**i** – 3**j**) N. Find **P**.

Add the forces:
$$-\mathbf{i} + 6\mathbf{j}$$
$$+\ \underline{3\mathbf{i} - 5\mathbf{j}}$$
$$2\mathbf{i} + \mathbf{j}$$

Subtract the total from the resultant:
$$7\mathbf{i} - 3\mathbf{j}$$
$$-\ \underline{2\mathbf{i} + \mathbf{j}}$$
$$5\mathbf{i} - 4\mathbf{j}$$

Answer: (5**i** – 4**j**) N

Exercise 19D:

1. The forces **P** N, (3**i** + **j**) N and (– 2**i** + 4**j**) N act on a body. The resultant of these forces is (2**i** + 3**j**) N. Find **P**.

2. The forces **P** N, (5**i** + 2**j**) N and (**i** – 6**j**) N act on a body. The resultant of these forces is 3**i** N. Find **P**.

3. The forces **P** N, (2**i** – 4**j**) N and (– **i** + 7**j**) N act on a body. The resultant of these forces is (2**i** + 8**j**) N. Find **P**.

4. The forces **P** N, (**i** + 3**j**) N and (2**i** – 4**j**) N act on a body. The resultant of these forces is (–**i** – 3**j**) N. Find **P**.

5. The forces **P** N, (4**i** – 2**j**) N and (–3**i** + **j**) N act on a body. The resultant of these forces is (4**i** + 7**j**) N. Find **P**.

6. The forces **P** N, (4**i** – 5**j**) N and (–2**i** + 6**j**) N act on a body. The resultant of these forces is 4**j** N. Find **P**.

Exercise 19D...

7. The forces **P** N, (−3**i** − 2**j**) N and (**i** − 4**j**) N act on a body. The resultant of these forces is (3**i** − 7**j**) N. Find **P**.

8. The forces **P** N, (4**i** − 3**j**) N and (−**i** − 2**j**) N act on a body. The resultant of these forces is (3**i** − 7**j**) N. Find **P**.

9. The forces **P** N, (−2**i** + **j**) N and (3**i** − 5**j**) N act on a body. The resultant of these forces is (5**i** − **j**) N. Find **P**.

10. The forces **P** N, (6**i** − 2**j**) N and (−4**i** + 5**j**) N act on a body. The resultant of these forces is (−**i** + 2**j**) N. Find **P**.

19.5 Newton's Law

Newton's second law can be written in vector form as:
$$\mathbf{F} = m\mathbf{a}$$
For example:

1. The forces (6**i** − 2**j**) N and (−4**i** + 5**j**) N act on a body of mass 5 kg. Find the acceleration.

To answer this question, you need to:
• Find the resultant force.
• Use **F** = m**a**.

Resultant:
$$\begin{array}{r} 6\mathbf{i} - 2\mathbf{j} \\ + \underline{-4\mathbf{i} + 5\mathbf{j}} \\ 2\mathbf{i} + 3\mathbf{j} \end{array}$$

Use:
$$\mathbf{F} = m\mathbf{a}$$
$$2\mathbf{i} + 3\mathbf{j} = 5\mathbf{a}$$

So
$$\mathbf{a} = \frac{1}{5}(2\mathbf{i} + 3\mathbf{j}) \text{ m/s}^2$$

2. The forces (3**i** − 5**j**) N and **P** N act on a body of mass 2kg. The acceleration is (2**i** + 5**j**) m/s². Find **P**.

To answer this question, you need to:
• Find the resultant force using **F** = m**a**.
• Subtract the known force from the resultant force.

Use
$$\mathbf{F} = m\mathbf{a}$$
$$\mathbf{F} = 2(2\mathbf{i} + 5\mathbf{j})$$
$$\mathbf{F} = 4\mathbf{i} + 10\mathbf{j}$$

Then subtract:
$$\begin{array}{r} 4\mathbf{i} + 10\mathbf{j} \\ - \underline{3\mathbf{i} - 5\mathbf{j}} \\ \mathbf{i} + 15\mathbf{j} \end{array}$$

Answer: (**i** + 15**j**) N

Exercise 19E:

1. The forces (3**i** − 2**j**) N and (2**i** − 13**j**) N act on a body of mass 5kg. Find the acceleration.

2. The forces (−5**i** − **j**) N and **P** N act on a body of mass 3 kg. The acceleration is (−2**i** + **j**) m/s². Find **P**.

3. The forces (5**i** − 2**j**) N and (−11**i** − 10**j**) N act on a body of mass 6kg. Find the acceleration.

4. The forces (30**i** − 15**j**) N and **P** N act on a body of mass 7 kg. The acceleration is (4**i** − 3**j**) m/s². Find **P**.

5. The forces (4**i** +7**j**) N and (−12**i** + 5**j**) N act on a body of mass 4 kg. Find the acceleration.

6. The forces (39**i** + 10**j**) N and **P** N act on a body of mass 9 kg. The acceleration is (4**i** + 2**j**) m/s². Find **P**.

7. The forces (2**i** − **j**) N and (−4**i** + 7**j**) N act on a body of mass 2 kg. Find the acceleration.

8. The forces (9**i** + 2**j**) N and **P** N act on a body of mass 3 kg. The acceleration is (2**i** − **j**) m/s². Find **P**.

9. The forces (4**i** − 9**j**) N and (5**i** + 15**j**) N act on a body of mass 3 kg. Find the acceleration.

10. The forces (−3**i** − 15**j**) N and **P** N act on a body of mass 5kg. The acceleration is (−**i** − 2**j**) m/s². Find **P**.

19.6 Constantly Accelerated Motion

You can use the following equations of motion written in vector form:
$$\mathbf{v} = \mathbf{u} + \mathbf{a}t$$
$$\mathbf{s} = \mathbf{u}t + \frac{1}{2}\mathbf{a}t^2$$
$$\mathbf{s} = \frac{1}{2}t(\mathbf{u} + \mathbf{v})$$

For example:

A body of mass 6kg moving with a velocity of (3**i** + 2**j**) m/s accelerates at (7**i** − 2**j**) m/s² for 3 s.
Find (i) the force producing the acceleration.
    (ii) the final velocity.
    (iii) the displacement after 3 s.

(i) Use
$$\mathbf{F} = m\mathbf{a}$$
$$\mathbf{F} = 6(7\mathbf{i} - 2\mathbf{j})$$
$$\mathbf{F} = (42\mathbf{i} - 12\mathbf{j}) \text{ N}$$

(ii) We know
$$\mathbf{u} = 3\mathbf{i} + 2\mathbf{j}$$
$$\mathbf{a} = 7\mathbf{i} - 2\mathbf{j}$$
$$t = 3$$

We want to find **v** = ?

So we use
$$\mathbf{v} = \mathbf{u} + \mathbf{a}t$$
$$\mathbf{v} = (3\mathbf{i} + 2\mathbf{j}) + 3(7\mathbf{i} - 2\mathbf{j})$$
$$\mathbf{v} = 3\mathbf{i} + 2\mathbf{j} + 21\mathbf{i} - 6\mathbf{j}$$
$$\mathbf{v} = (24\mathbf{i} - 4\mathbf{j}) \text{ m/s}$$

(iii) We know
$$\mathbf{u} = 3\mathbf{i} + 2\mathbf{j}$$
$$\mathbf{a} = 7\mathbf{i} - 2\mathbf{j}$$
$$t = 3$$

We want to find $\mathbf{s} = ?$

So we use
$$\mathbf{s} = \mathbf{u}t + \frac{1}{2}\mathbf{a}t^2$$
$$\mathbf{s} = 3(3\mathbf{i} + 2\mathbf{j}) + \frac{1}{2}(9)(7\mathbf{i} - 2\mathbf{j})$$
$$\mathbf{s} = 9\mathbf{i} + 6\mathbf{j} + 4.5(7\mathbf{i} - 2\mathbf{j})$$
$$\mathbf{s} = 9\mathbf{i} + 6\mathbf{j} + 31.5\mathbf{i} - 9\mathbf{j}$$
$$\mathbf{s} = (40.5\mathbf{i} - 3\mathbf{j}) \text{ m}$$

Exercise 19F:

1. A body of mass 4 kg moving with a velocity of $(3\mathbf{i} - 2\mathbf{j})$ m/s accelerates at $(\mathbf{i} - 3\mathbf{j})$ m/s^2 for 3 s.
   Find (i) the force producing the acceleration.
   (ii) the final velocity. (iii) the displacement after 3 s.

2. A body of mass 5 kg moving with a velocity of $(-\mathbf{i} + 2\mathbf{j})$ m/s is acted upon by a force of $(10\mathbf{i} + 10\mathbf{j})$ N and accelerates for 2 s.
   Find (i) the acceleration. (ii) the final velocity.
   (iii) the displacement after 2 s.

3. A body of mass 2 kg accelerates at $(\mathbf{i} + 3\mathbf{j})$ m/s^2 for 6 s reaching a final velocity of $(4\mathbf{i} + 14\mathbf{j})$ m/s.
   Find (i) the force producing the acceleration.
   (ii) the inital velocity. (iii) the displacement after 6 s.

4. A body of mass 8 kg moving with a velocity of $(4\mathbf{i} - 3\mathbf{j})$ m/s is acted upon by a force of $-60\mathbf{j}$ N and accelerates for 4 s.
   Find (i) the acceleration. (ii) the final velocity.
   (iii) the displacement after 4 s.

5. A body of mass 6 kg accelerates at $(-2\mathbf{i} - 5\mathbf{j})$ m/s^2 for 3 s reaching a final velocity of $(-3\mathbf{i} - 11\mathbf{j})$ m/s.
   Find (i) the force producing the acceleration.
   (ii) the inital velocity. (iii) the displacement after 3 s.

6. A body of mass 3 kg moving with a velocity of $(-2\mathbf{i} - 5\mathbf{j})$ m/s is acted upon by a force of $(7.5\mathbf{i} + 37.5\mathbf{j})$ N and accelerates for 5 s.
   Find (i) the acceleration. (ii) the final velocity.
   (iii) the displacement after 5 s.

7. A body of mass 7 kg accelerates at $(-2\mathbf{i} + \mathbf{j})$ m/s^2 for 4 s and travels a displacement of $-4\mathbf{j}$ m.
   Find (i) the force producing the acceleration.
   (ii) the initial velocity. (iii) the final velocity.

Exercise 19F...

8. A body of mass 10 kg moving with a velocity of $(-\mathbf{i} + 3\mathbf{j})$ m/s is acted upon by a force of $(40\mathbf{i} + 20\mathbf{j})$ N and accelerates for 2 s.
   Find (i) the acceleration.
   (ii) the final velocity.
   (iii) the displacement after 2 s.

9. A body of mass 4 kg accelerates at $(-2\mathbf{i} + 4\mathbf{j})$ m/s^2 for 2 s and travels a displacement of $(6\mathbf{i} + 4\mathbf{j})$ m.
   Find (i) the force producing the acceleration.
   (ii) the initial velocity.
   (iii) the final velocity.

10. A body of mass 2 kg is acted upon by a force of $(-60\mathbf{i} + 36\mathbf{j})$ N and accelerates for 6 s reaching a final velocity of $(-8\mathbf{i} + 12\mathbf{j})$ m/s.
    Find (i) the acceleration.
    (ii) the initial velocity.
    (iii) the displacement after 6 s.

19.7 Further Problems

The same methods can be used to solve more complex problems. For example:

A body is moving parallel to the vector $3\mathbf{i} - 4\mathbf{j}$ with a speed of 65 m/s. Find its velocity.

To answer this question, you need to:
• Find the magnitude of the parallel vector.
• Work out how many times the speed is greater than this magnitude.

The magnitude of the parallel vector is:
$$R = \sqrt{3^2 + (-4)^2}$$
$$R = \sqrt{25}$$
$$R = 5$$

Next, find how many times the speed is greater than this magnitude:
$$\frac{65}{5} = 13$$

So the velocity is 13 times the vector ie:
$$13(3\mathbf{i} - 4\mathbf{j}) = (39\mathbf{i} - 52\mathbf{j}) \text{ m/s}$$

Exercise 19G:

1. A body is moving parallel to the vector $5\mathbf{i} - 12\mathbf{j}$ with a speed of 26 m/s. Find its velocity.

2. A body is moving parallel to the vector $-3\mathbf{i} + 4\mathbf{j}$ with a speed of 30 m/s. Find its velocity.

3. A body is moving parallel to the vector 15**i** + 8**j** with a speed of 51 m/s. Find its velocity.

4. A body is moving parallel to the vector –7**i** – 24**j** with a speed of 75 m/s. Find its velocity.

5. A body is moving parallel to the vector 4**i** + 3**j** with a speed of 45 m/s. Find its velocity.

6. A body is moving parallel to the vector –12**i** + 5**j** with a speed of 39 m/s. Find its velocity.

CHAPTER 20: FRICTION

20.1 The Concept of Friction

Friction is a **force** which always acts on rough surfaces. The frictional force is proportional to the normal reaction between a body and a surface (ie the perpendicular reaction).

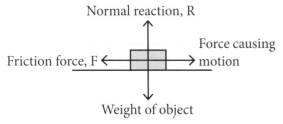

Normal reaction, R

Friction force, F

Force causing motion

Weight of object

This can be written as follows:

When a body moves or is just on the point of moving
$$F \leq \mu R$$
where μ is the coefficient of friction

Friction always opposes the direction of motion in which a body is moving or the direction of motion in which it would move. Friction can only prevent motion up to a limiting value after which motion will take place.

20.2 Bodies Just About To Move – Horizontal Motion

Remember, friction always acts opposite to the direction in which the body would move. The rules for solving this type of problem are:
• Resolve forces **horizontally** and **vertically**.
• Equate the resolved forces in each direction (because you know that the block is just about to move).
For example:

1. A block mass 6 kg rests on a rough horizontal plane. A horizontal force of 32 N will just cause the block to start to move. Find the coefficient of friction.

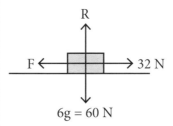

$$6g = 60 \text{ N}$$

Resolving horizontally:
$$F = 32$$

Resolving vertically:
$$R = 6g = 60$$

Now find μ:
$$F = \mu R$$
$$\mu = \frac{F}{R}$$
$$\mu = \frac{32}{60} = 0.533$$

2. A block mass 7.2 kg rests on a rough horizontal plane. The coefficient of friction is 0.8. Find the least horizontal force needed to move it.

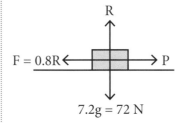

$$7.2g = 72 \text{ N}$$

Let the force be called P.

Resolving vertically:
$$R = 7.2g = 72$$

Resolving horizontally:
$$P = 0.8R$$
$$= 0.8 \times 72$$
$$= 57.6 \text{ N}$$

Exercise 20A:

1. A block of mass 4 kg rests on a rough horizontal plane. A horizontal force of 25 N will just cause the block to start to move. Find the coefficient of friction.

2. A block of mass 2.8 kg rests on a rough horizontal plane. A horizontal force of 18 N will just cause the block to start to move. Find the coefficient of friction.

3. A block of mass 24 kg rests on a rough horizontal plane. The coefficient of friction is 0.76. Find the least horizontal force needed to move it.

4. A block of mass 5.4 kg rests on a rough horizontal plane. The coefficient of friction is 0.64. Find the least horizontal force needed to move it.

5. A block of mass M kg rests on a rough horizontal plane. The coefficient of friction is 0.85. The least horizontal force needed to move it is 28 N. Find M.

6. A block of mass M kg rests on a rough horizontal plane. The coefficient of friction is 0.72. The least horizontal force needed to move it is 12 N. Find M.

7. A block of mass 4.8 kg rests on a rough horizontal plane. A horizontal force of 25 N will just cause the block to start to move. Find the coefficient of friction.

8. A block mass 12 kg rests on a rough horizontal plane. The coefficient of friction is 0.56. Find the least horizontal force needed to move it.

9. A block of mass M kg rests on a rough horizontal plane. The coefficient of friction is 0.86. The least horizontal force needed to move it is 32N. Find M.

10. A block of mass 28 kg rests on a rough horizontal plane. The coefficient of friction is 0.68. Find the least horizontal force needed to move it.

20.3 Bodies Just About To Move – Inclined Plane

When the block lies on an inclined plane, the rules are similar:
• Resolve forces **along** the plane and **perpendicular** to the plane.
• Equate the resolved forces in each direction (because you know that the block is just about to move).
For example:

1. A block of mass 4 kg rests on a rough plane inclined at 25° to the horizontal. It is just about to move. Find the coefficient of friction.

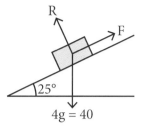

If it is just about to move then friction will act **up** the slope. Thus:

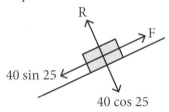

Resolving along the plane:
$$F = 40 \sin 25$$
Resolving perpendicular to the plane:
$$R = 40 \cos 25$$
Now find μ:
$$\mu = \frac{F}{R}$$
$$\mu = \frac{40 \sin 25}{40 \cos 25}$$
$$\mu = 0.466$$

....

2. A block of mass 5 kg rests on a rough plane inclined at 30° to the horizontal. The coefficient of friction is 0.3.

Find the least force needed to
(i) move it up the plane
(ii) stop it falling down the plane.

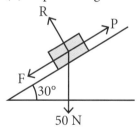

(i) If it is to move **up** the plane then friction will act **down**. Thus:

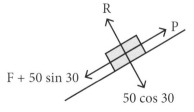

Let the force be called P.
Resolving perpendicular to the plane:
$$R = 50 \cos 30$$
$$R = 43.3 \text{ N}$$

Resolving along the plane:
$$P = F + 50 \sin 30$$
$$P = F + 25$$
But $\quad\quad F = \mu R$
So $\quad\quad F = 0.3 \times 43.3$
$$F = 12.99$$
So $\quad\quad P = 12.99 + 25$
$$P = 37.99 \text{ N}$$

(ii) If it is to stop it falling **down** the plane then friction will act **up**.

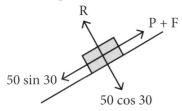

Let the force be called P.
Resolving perpendicular to the plane:
$$R = 50 \cos 30$$
$$= 43.3 \text{ N}$$
Resolving along the plane:
$$P + F = 50 \sin 30$$
$$P + F = 25$$
But $\quad\quad F = \mu R$
So $\quad\quad F = 0.3 \times 43.3$
$$= 12.99 \text{ N}$$
So $\quad P + 12.99 = 25$
$$P = 25 - 12.99$$
$$P = 12.01 \text{ N}$$

Exercise 20B:

1. A block of mass 8 kg rests on a rough plane inclined at 20° to the horizontal. It is just about to move. Find the coefficient of friction.

2. A block of mass 12 kg rests on a rough plane inclined at 42° to the horizontal. It is just about to move. Find the coefficient of friction.

3. A block of mass 6 kg rests on a rough plane inclined at 24° to the horizontal. The coefficient of friction is 0.16. Find the least force needed to
   (i) move it up the plane
   (ii) stop it falling down the plane.

4. A block of mass 8 kg rests on a rough plane inclined at 32° to the horizontal. The coefficient of friction is 0.24. Find the least force needed to
   (i) move it up the plane
   (ii) stop it falling down the plane.

Exercise 20B...

5. A block of mass 5.6 kg rests on a rough plane inclined at 15° to the horizontal. The coefficient of friction is 0.17. Find the least force needed to
   (i) move it up the plane
   (ii) stop it falling down the plane.

6. A block of mass 2.8 kg rests on a rough plane inclined at 35° to the horizontal. The coefficient of friction is 0.41. Find the least force needed to
   (i) move it up the plane
   (ii) stop it falling down the plane.

7. A block of mass 14 kg rests on a rough plane inclined at 22° to the horizontal. The coefficient of friction is 0.36. Find the least force needed to
   (i) move it up the plane
   (ii) stop it falling down the plane.

8. A block of mass 3.7 kg rests on a rough plane inclined at 26° to the horizontal. The coefficient of friction is 0.32. Find the least force needed to
   (i) move it up the plane
   (ii) stop it falling down the plane.

9. A block of mass 4.3 kg rests on a rough plane inclined at 14° to the horizontal. The coefficient of friction is 0.24. Find the least force needed to
   (i) move it up the plane
   (ii) stop it falling down the plane.

10. A block mass 7.8 kg rests on a rough plane inclined at 34° to the horizontal. The coefficient of friction is 0.54. Find the least force needed to
   (i) move it up the plane
   (ii) stop it falling down the plane.

20.4 Bodies Moving – Horizontal Motion

Friction always acts opposite to the direction in which the body moves. For example:

A block of mass 7.6 kg is initially at rest. It is pulled along a rough horizontal plane by a horizontal force of 58 N which will just cause the block to start to move. The coefficient of friction is 0.6. Find
(i) the acceleration
(ii) the time it takes to reach a velocity of 4 m/s.

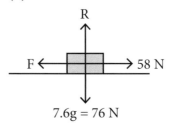

The rules are:
• Resolve forces horizontally and vertically.
• Equate the resolved forces vertically.
• Find the resultant horizontal force.
• Use F = ma

(i) Resolving vertically:
$$R = 7.6g = 76 \text{ N}$$
Resultant horizontal force:
$$= 58 - F$$
Since F = μR:
$$F = 0.6 \times 76$$
$$= 45.6 \text{ N}$$
So the resultant horizontal force
$$= 58 - 45.6$$
$$= 12.4 \text{ N}$$
Find the acceleration:
$$F = ma$$
$$12.4 = 7.6a$$
$$a = \frac{12.4}{7.6}$$
$$= 1.63 \text{ m/s}^2$$
(ii) We know   u = 0
   a = 1.63
   v = 4
We want to find t = ?
So we use   v = u + at
$$4 = 0 + 1.63t$$
$$t = \frac{4}{1.63}$$
$$t = 2.45 \text{ s}$$

Exercise 20C:

1. A block of mass 3.4 kg is initially at rest. It is pulled along a rough horizontal plane by a horizontal force of 42 N will just cause the block to start to move. The coefficient of friction is 0.27. Find:
(i) the acceleration
(ii) the time it takes to reach a velocity of 2.5 m/s.

2. A block of mass 7.2 kg accelerates from 2 m/s to 26 m/s in 8 s on a rough horizontal plane when acted upon by a horizontal force of 48 N. Find:
(i) the coefficient of friction
(ii) the distance travelled.

3. A block of mass 6 kg is initially at rest. It is pulled along a rough horizontal plane by a horizontal force of 72 N which will just cause the block to start to move. The coefficient of friction is 0.48. Find:
(i) the acceleration
(ii) the time it takes to travel 5.4 m.

Exercise 20C...

4. A block of mass 2.8 kg accelerates from 4 m/s and travels 32 m in 4 s on a rough horizontal plane when acted upon by a horizontal force of 26 N. Find:
(i) the coefficient of friction (ii) the final velocity.

5. A block of mass 24 kg is initially at rest. It is pulled along a rough horizontal plane by a horizontal force of 176 N which will just cause the block to start to move. The coefficient of friction is 0.58. Find:
(i) the acceleration (ii) the distance travelled in 5 s.

6. A block of mass 4.3 kg accelerates from 6 m/s to 56 m/s in travelling 310 m on a rough horizontal plane when acted upon by a horizontal force of 52 N. Find:
(i) the coefficient of friction (ii) the time taken.

7. A block of mass 4.8 kg is initially at rest. It is pulled along a rough horizontal plane by a horizontal force of 37 N which will just cause the block to start to move. The coefficient of friction is 0.72. Find:
(i) the acceleration
(ii) the velocity when it has travelled 8.2 m.

8. A block of mass 5.8 kg accelerates reaching a velocity of 27 m/s in 6 s while travelling 90 m on a rough horizontal plane when acted upon by a horizontal force of 68 N. Find:
(i) the coefficient of friction (ii) the initial velocity.

9. A block mass 7.6 kg is initially at rest. It is pulled along a rough horizontal plane by a horizontal force of 58 N which will just cause the block to start to move. The coefficient of friction is 0.6. Find:
(i) the acceleration (ii) the velocity after 4 s.

10. A block of mass 8 kg is initially at rest. It is pulled along a rough horizontal plane by a horizontal force of 44 N which will just cause the block to start to move. The coefficient of friction is 0.4. Find:
(i) the acceleration
(ii) the distance travelled in 5 s.

11. A block of mass 3.7 kg accelerates from 7 m/s and travels 504 m in 12 s on a rough horizontal plane when acted upon by a horizontal force of 40 N. Find:
(i) the coefficient of friction (ii) the final velocity.

20.5 Bodies Moving – Inclined Plane

The same principles can be applied to bodies moving on inclined planes. For example:

A block of mass 7.2 kg rests on a rough plane inclined at 36° to the horizontal. A force of 64 N parallel to, and acting up, the plane is applied to the body and it begins to accelerate reaching 6 m/s in 3 s. Find the coefficient of friction.

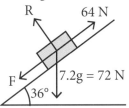

You must first find the acceleration as follows.

We know	$u = 0$
	$v = 6$
	$t = 3$

We want to find $a = ?$

So we use $v = u + at$

$6 = 0 + 3a$

$a = 2 \text{ m/s}^2$

You must now resolve the forces as shown:

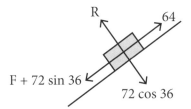

Resolving perpendicular to the plane:

$R = 72 \cos 36$

$= 58.2$

Resultant force along the plane

$= 64 - F - 72\sin36$

Using $F = ma$

$64 - F - 72 \sin 36 = 7.2 \times 2$

$64 - F - 42.3 = 14.4$

$21.7 - F = 14.4$

$21.7 - 14.4 = F$

$F = 7.3 \text{ N}$

Now find μ: $\mu = \dfrac{F}{R}$

$\mu = \dfrac{7.3}{58.2} = 0.125$

Exercise 20D:

1. A block of mass 6 kg rests on a rough plane inclined at 40° to the horizontal. A force of 84 N parallel to, and acting up, the plane is applied to the body and it begins to accelerate travelling 5 m in 2.5 s. Find the coefficient of friction.

2. A block of mass 6.2 kg rests on a rough plane inclined at 36° to the horizontal. A force of 52 N parallel to, and acting up, the plane is applied to the body and it begins to accelerate reaching 6 m/s in 5 s. Find the coefficient of friction.

3. A block of mass 5.8 kg rests on a rough plane inclined at 24° to the horizontal. A force of P N parallel to, and acting up, the plane is applied to the body and it begins to accelerate reaching 3 m/s in travelling 2.25 m. The coefficient of friction is 0.68. Find P.

4. A block of mass 3.9 kg rests on a rough plane inclined at 35° to the horizontal. A force of P N parallel to, and acting up, the plane is applied to the body and it begins to accelerate travelling 5 m in 2 s. The coefficient of friction is 0.76. Find P.

5. A block of mass 4.5 kg rests on a rough plane inclined at 40° to the horizontal. A force of 54 N parallel to, and acting up, the plane is applied to the body and it begins to accelerate reaching 9.6 m/s in 3 s. Find the coefficient of friction.

6. A block of mass 9 kg rests on a rough plane inclined at 56° to the horizontal. A force of 95 N parallel to, and acting up, the plane is applied to the body and it begins to accelerate reaching 2.1 m/s in travelling 1.575 m. Find the coefficient of friction.

7. A block of mass 7 kg rests on a rough plane inclined at 25° to the horizontal. A force of P N parallel to, and acting up, the plane is applied to the body and it begins to accelerate travelling 3 m in 2 s. The coefficient of friction is 0.46. Find P.

8. A block of mass 8 kg rests on a rough plane inclined at 42° to the horizontal. A force of 78 N parallel to, and acting up, the plane is applied to the body and it begins to accelerate reaching 2.8 m/s in 1.4 s. Find the coefficient of friction.

9. A block of mass 5.6 kg rests on a rough plane inclined at 26° to the horizontal. A force of P N parallel to, and acting up, the plane is applied to the body and it begins to accelerate reaching 3 m/s in travelling 7.5 m. The coefficient of friction is 0.44. Find P.

10. A block of mass 4.8 kg rests on a rough plane inclined at 33° to the horizontal. A force of 49 N parallel to, and acting up, the plane is applied to the body and it begins to accelerate reaching travelling 80 cm in 2 s. Find the coefficient of friction.

CONNECTED BODIES

21.1 The Concept of Connected Bodies

Sometimes you have to deal with problems of forces acting on two connected bodies. The rules for solving this type of problem are:

- Consider the two bodies separately
- Use F = ma

For example:

1. A van of mass 1100 kg tows a trailer of mass 560 kg by means of a light horizontal tow bar. The tractive force produced by the van's engine is 3960 N. The van and trailer travel along a straight horizontal road as shown below and accelerate at 1.6 m/s². The resistance to motion of the van is 950 N. Find:

(i) the resistance to motion of the trailer
(ii) the magnitude of the tension in the tow bar.

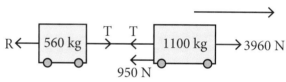

Let T = the magnitude of the tension in the tow bar
and R = the resistance to motion of the trailer
First consider the van:
$$F = ma$$
$$3960 - 950 - T = 1100 \times 1.6$$
$$3010 - T = 1760$$
$$3010 - 1760 = T$$
$$T = 1250 \text{ N}$$
Next consider the trailer:
$$F = ma$$
$$T - R = 560 \times 1.6$$
$$1250 - R = 560 \times 1.6$$
$$1250 - R = 896$$
$$1250 - 896 = R$$
$$R = 354 \text{ N}$$
Answers: (i) 354 N (ii) 1250 N

2. A van of mass 1050 kg tows a trailer of mass 520 kg by means of a light horizontal tow bar. The tractive force produced by the van's engine is 3750 N. The van and trailer travel along a straight horizontal road. The resistance to motion of the van is 860 N and the resistance to motion of the trailer is 425 N. Find:

(i) the acceleration
(ii) the magnitude of the tension in the tow bar.

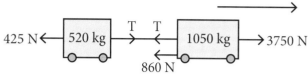

Let T = the magnitude of the tension in the tow bar
and a = the acceleration
First consider the van:
$$F = ma$$
$$3750 - 860 - T = 1050a$$
$$2890 - T = 1050a$$
Next consider the trailer:
$$F = ma$$
$$T - 425 = 520a$$
You can add these equations to get rid of the T:
$$2890 - T = 1050a$$
$$+ \quad \underline{T - 425 = 520a}$$
$$2465 = 1570a$$
$$a = \frac{2465}{1570}$$
$$a = 1.57 \text{ m/s}^2$$
Substituting a into either equation will give T:
$$T - 425 = 520a$$
$$T - 425 = 520 \times 1.57$$
$$T - 425 = 816.4$$
$$T = 816.4 + 425$$
$$T = 1241.4 \text{ N}$$
Answers: (i) 1.57 m/s² (ii) 1241.4 N

Exercise 21A:

1. A van of mass 1060 kg tows a trailer of mass M kg by means of a light horizontal tow bar. The tractive force produced by the van's engine is 3160 N. The van and trailer travel along a straight horizontal road. The resistance to motion of the van is 925 N and the resistance to motion of the trailer is 435 N. The magnitude of the tension in the tow bar is 980 N. Find:
(i) the acceleration
(ii) M.

2. A van of mass 1160 kg tows a trailer of mass 480 kg by means of a light horizontal tow bar. The tractive force produced by the van's engine is 3680 N. The van and trailer travel along a straight horizontal road. The resistance to motion of the van is 925 N and the resistance to motion of the trailer is 385 N. Find:
(i) the acceleration
(ii) the magnitude of the tension in the tow bar.

3. A van of mass 1080 kg tows a trailer of mass 450 kg by means of a light horizontal tow bar. The tractive force produced by the van's engine is 3750 N. The van and trailer travel along a straight horizontal road and accelerate at 1.3 m/s². The resistance to motion of the van is 924 N. Find:
   (i) the resistance to motion of the trailer
   (ii) the magnitude of the tension in the tow bar.

4. A van of mass 1120 kg tows a trailer of mass 385 kg by means of a light horizontal tow bar. The tractive force produced by the van's engine is 3880 N. The van and trailer travel along a straight horizontal road and accelerate at 1.25 m/s². The resistance to motion of the trailer is 575 N. Find:
   (i) the resistance to motion of the van
   (ii) the magnitude of the tension in the tow bar.

5. A van of mass 1200 kg tows a trailer of mass M kg by means of a light horizontal tow bar. The tractive force produced by the van's engine is PN. The van and trailer travel along a straight horizontal road and accelerate at 1.4 m/s². The resistance to motion of the van is 845 N and the resistance to motion of the trailer is 620 N. The magnitude of the tension in the tow bar is 975 N. Find:
   (i) P
   (ii) M.

6. A van of mass M kg tows a trailer of mass m kg by means of a light horizontal tow bar. The tractive force produced by the van's engine is 3440 N. The van and trailer travel along a straight horizontal road and accelerate at 1.15 m/s². The resistance to motion of the van is 920 N and the resistance to motion of the trailer is 546 N. The magnitude of the tension in the tow bar is 864 N. Find:
   (i) M
   (ii) m

7. A van of mass 980 kg tows a trailer of mass 420 kg by means of a light horizontal tow bar. The tractive force produced by the van's engine is 3050 N. The van and trailer travel along a straight horizontal road. The resistance to motion of the van is 825 N. The magnitude of the tension in the tow bar is 910 N. Find:
   (i) the acceleration
   (ii) the resistance to motion of the van.

8. A van of mass M kg tows a trailer of mass 375 kg by means of a light horizontal tow bar. The tractive force produced by the van's engine is 3220 N. The van and trailer travel along a straight horizontal road and accelerate at 1.08 m/s². The resistance to motion of the van is 810 N and the resistance to motion of the trailer is 380 N. Find:
   (i) M
   (ii) the magnitude of the tension in the tow bar.

9. A body of mass 8 kg rests on a rough horizontal plane. The coefficient of friction is $\frac{1}{3}$. It is just about to move when a force of P N is applied at 20° to the horizontal. Find the value of P.

10. A body of mass 8.4 kg is pulled along a rough horizontal plane by a force of 48 N acting at 35° to the horizontal at an acceleration of 2 m/s². Find the coefficient of friction.

11. A body of mass M kg rests on a rough horizontal plane. The coefficient of friction is $\frac{2}{5}$. It is just about to move when a force of 35 N is applied at 16° to the horizontal. Find the value of M.

12. A body of mass 6.5 kg is pulled along a rough horizontal plane by a force of 56 N acting at 24° to the horizontal at an acceleration of **a** m/s². The coefficient of friction is 0.7. Find the value of **a**.

21.2 Resistance Per Mass

Sometimes the resistance to the motion is given **per mass**. To find the **total** resistance you would then multiply this figure by the mass. If the resistance to the motion of a car of mass 960 kg was given as **0.642 N per kg of mass** then the total resistance would be 960 × 0.642 = 616.32 N.

For example:
A van of mass 1125 kg tows a trailer of mass 540 kg by means of a light horizontal tow bar. The tractive force produced by the van's engine is 3250 N. The van and trailer travel along a straight horizontal road. The resistance to motion of the van is 0.86 N per kg of mass and the resistance to motion of the trailer is 0.82 N per kg of mass. Find:
(i) the acceleration
(ii) the magnitude of the tension in the tow bar

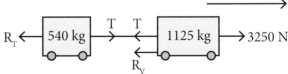

Let T = the magnitude of the tension in the tow bar and a = the acceleration.

First consider the van:

Total resistance R_V is:

$0.86 \times 1125 = 967.5$ N

Using $\qquad$ F = ma

$3250 - 967.5 - T = 1125a$

$2282.5 - T = 1125a$

Next consider the trailer:

Total resistance R_T is:

$0.82 \times 540 = 442.8$ N

Using $\qquad$ F = ma

$T - 442.8 = 540a$

You can add these equations to get rid of the T:

$2282.5 - T = 1125a$

$+ \quad \underline{T - 442.8 = 540a}$

$1839.7 = 1665a$

$a = \dfrac{1839.7}{1665}$

$a = 1.1$ m/s^2

Substituting a into either equation will give T:

$T - 442.8 = 540a$

$T - 442.8 = 540 \times 1.1$

$T - 442.8 = 594$

$T = 594 + 442.8$

$T = 1036.8$ N

Answers (i) 1.1 m/s^2 (ii) 1036.8 N

Exercise 21B:

1. A van of mass 1140 kg tows a trailer of mass 540 kg by means of a light horizontal tow bar. The tractive force produced by the van's engine is 3460 N. The van and trailer travel along a straight horizontal road. The resistance to motion of the van is 0.88 N per kg of mass and the resistance to motion of the trailer is 0.83 N per kg of mass. Find:
   (i) the acceleration
   (ii) the magnitude of the tension in the tow bar.

2. A van of mass 1025 kg tows a trailer of mass 496 kg by means of a light horizontal tow bar. The tractive force produced by the van's engine is 3375 N. The van and trailer travel along a straight horizontal road. The resistance to motion of the van is 0.92 N per kg of mass and the resistance to motion of the trailer is 0.87 N per kg of mass. Find:
   (i) the acceleration
   (ii) the magnitude of the tension in the tow bar.

3. A van of mass 986 kg tows a trailer of mass 424 kg by means of a light horizontal tow bar. The tractive force produced by the van's engine is 3215 N. The van and trailer travel along a straight horizontal road and accelerate at 1.2 m/s^2. The resistance to motion of the van is 0.92 N per kg of mass. Find:

Exercise 21B...

   (i) the magnitude of the tension in the tow bar
   (ii) the resistance to motion of the trailer per kg of mass.

4. A van of mass 1024 kg tows a trailer of mass 414 kg by means of a light horizontal tow bar. The tractive force produced by the van's engine is 3540 N. The van and trailer travel along a straight horizontal road and accelerate at 0.96 m/s^2. The resistance to motion of the trailer is 1.12 N per kg of mass. Find:
   (i) the magnitude of the tension in the tow bar
   (ii) the resistance to motion of the van per kg of mass.

5. A van of mass 1156 kg tows a trailer of mass M kg by means of a light horizontal tow bar. The tractive force produced by the van's engine is P N. The van and trailer travel along a straight horizontal road and accelerate at 1.28 m/s^2. The resistance to motion of the van is 0.84 N per kg of mass and the resistance to motion of the trailer is 0.8 N per kg of mass. The magnitude of the tension in the tow bar is 920 N. Find:
   (i) P (ii) M.

6. A van of mass M kg tows a trailer of mass m kg by means of a light horizontal tow bar. The tractive force produced by the van's engine is 3356 N. The van and trailer travel along a straight horizontal road and accelerate at 1.08 m/s^2. The resistance to motion of the van is 1.05 N per kg of mass and the resistance to motion of the trailer is 0.96 N per kg of mass. The magnitude of the tension in the tow bar is 875 N. Find:
   (i) M (ii) m.

7. A van of mass 976 kg tows a trailer of mass 448 kg by means of a light horizontal tow bar. The tractive force produced by the van's engine is 3164 N. The van and trailer travel along a straight horizontal road. The resistance to motion of the van is 1.02 N per kg of mass. The magnitude of the tension in the tow bar is 920 N. Find:
   (i) the acceleration
   (ii) the resistance to motion of the trailer per kg of mass.

8. A van of mass M kg tows a trailer of mass 386 kg by means of a light horizontal tow bar. The tractive force produced by the van's engine is 3274 N. The van and trailer travel along a straight horizontal road and accelerate at 1.12 m/s^2. The resistance to motion of the van is 0.86 N per kg of mass and the

Exercise 21B...

resistance to motion of the trailer is 0.82 N per kg of mass. Find:
(i) M
(ii) the magnitude of the tension in the tow bar.

21.3 Resistance Per Mass – Harder Questions

Often you are asked more difficult questions involving resistance per mass. For example:

A car of mass 1400 kg tows a trailer of mass 600 kg at a constant speed of 14 m/s. The resistance to motion of the car is 1.3 N per kg of mass and the resistance to motion of the trailer is 0.9 N per kg of mass. Find:
(a) the tractive force of the car's engine
(b) the magnitude of the tension in the tow bar
(c) After travelling for a short time, the car accelerates at 0.2 m/s². Find:
  (i) the new tractive force of the engine
  (ii) the new magnitude of the tension in the tow bar
(d) If the tow bar breaks, find the deceleration of the trailer.

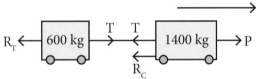

Let the tractive force = P
and T = the magnitude of the tension in the tow bar

(a) The car (and thus the trailer) are moving at a constant speed. So you must equate the forces.
Total resistance of the car R_C is:
$$1.3 \times 1400 = 1820 \text{ N}$$
Total resistance of the trailer R_T is:
$$0.9 \times 600 = 540 \text{ N}$$
You should consider the whole system, as the tensions will cancel each other out. Since the system is travelling at a constant speed, the tractive force of the engine is equal to the sum of the resistances of the car and trailer:
$$P = 540 + 1820$$
$$P = 2360 \text{ N}$$

(b) Since the system is travelling at a constant speed, the tension must be equal to the resistance of the trailer.
$$T = 540 \text{ N from part (a)}$$

(c) In this part of the question, the car is accelerating.
So you must use F = ma

First consider the car:
Using   F = ma
$$P - (1820 + T) = 1400 \times 0.2$$
$$P - 1820 - T = 280$$
$$P - T = 2100$$
Then consider the trailer:
Using   F = ma
$$T - 540 = 600 \times 0.2$$
$$T - 540 = 120$$
$$T = 660 \text{ N}$$
Substituting:
$$P - T = 2100$$
$$P - 660 = 2100$$
$$P = 2760 \text{ N}$$
So the answers are (i) 2760 N (ii) 660 N

(d) If the tow bar breaks then the only force acting on the trailer is –540 N which will slow it down. (The force is negative as acts opposite to the direction of travel.)
Using   F = ma
$$-540 = 600a$$
$$a = \frac{-540}{600}$$
$$a = -0.9 \text{ m/s}^2$$
Be careful: the question asked for the **de**celeration.
So the deceleration is 0.9 m/s²

Exercise 21C:

1. A car of mass 1500 kg tows a trailer of mass 450 kg at a steady speed. The resistance to motion of the car is 1.25 N per kg of mass and the resistance to motion of the trailer is 0.84 N per kg of mass. Find:
(a) the tractive force of the car's engine
(b) the magnitude of the tension in the tow bar
(c) After travelling for a short time, the car accelerates at 0.15 m/s². Find:
(i) the new tractive force of the engine
(ii) the new magnitude of the tension in the tow bar
(d) If the tow bar breaks find the deceleration of the trailer.

2. A car of mass 1350 kg tows a trailer of mass 540 kg at a constant speed. The resistance to motion of the car is 1.4 N per kg of mass and the resistance to motion of the trailer is 0.95 N per kg of mass. Find:
(a) the tractive force of the car's engine
(b) the magnitude of the tension in the tow bar
(c) After travelling for a short time, the car accelerates at 0.3 m/s². Find:
(i) the new tractive force of the engine
(ii) the new magnitude of the tension in the tow bar
(d) If the tow bar breaks find the deceleration of the trailer.

Exercise 21C...

3. A car of mass 1200 kg tows a caravan of mass 1500 kg at a steady speed. The resistance to motion of the car is 1.6 N per kg of mass and the resistance to motion of the caravan is 1.8 N per kg of mass. Find:
(a) the tractive force of the car's engine
(b) the magnitude of the tension in the tow bar
(c) After travelling for a short time, the car accelerates at 0.28 m/s². Find:
(i) the new tractive force of the engine
(ii) the new magnitude of the tension in the tow bar
(d) If the tow bar breaks find the deceleration of the trailer.

4. A car of mass 1100 kg tows a caravan of mass 1400 kg at an acceleration of 0.4 m/s². The resistance to motion of the car is 1.3 N per kg of mass and the resistance to motion of the caravan is 1.6 N per kg of mass. The car and caravan start from rest. Find:
(a) the tractive force of the car's engine
(b) the magnitude of the tension in the tow bar
(c) the distance travelled in 6 s
(d) the velocity of the car after it has travelled 5 m. As soon as the car has travelled 5 m the tow bar breaks.
(e) Find the additional distance the caravan travels before it comes to rest.

5. A van of mass 1170 kg tows a trailer of mass 520 kg by means of a light horizontal tow bar. The tractive force produced by the van's engine is 3756 N. The van and trailer accelerate uniformly from rest to 20 m/s in 16 s.
(i) Find the acceleration
The resistance to motion of the van is 0.78 N per kg of mass. Find:
(ii) the total resistance to the motion of the van and trailer
(iii) the resistance to the motion of the van
(iv) the resistance to the motion of the trailer
(v) the magnitude of the tension in the tow bar
The van and trailer then travel at a constant speed of 20 m/s and after some time the tow bar breaks.
(iv) Find the time before the trailer comes to rest.

6. A van of mass 1156 kg tows a trailer of mass 480 kg by means of a light horizontal tow bar. The tractive force produced by the van's engine is P N. The van and trailer accelerate uniformly from rest to 6 m/s in travelling 11.25m.
(i) Find the acceleration
The resistance to motion of the van is 0.92 N per kg of mass and the resistance to motion of the trailer is 0.85 N per kg of mass. Find:
(ii) the total resistance to the motion of the van and trailer
(iii) the magnitude of the tension in the tow bar
(iv) P
The van and trailer then travel at a constant speed of 6 m/s and after some time the tow bar breaks. Find:
(v) the time before the trailer comes to rest
(vi) the additional distance travelled before the trailer comes to rest.

21.4 Pulleys

Problems can also involve bodies that are connected by strings that pass over a pulley. In this case, the tensions in the connecting string act in the same direction. For example:

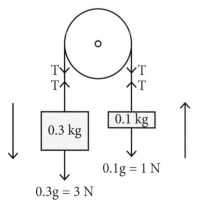

Two particles of mass 0.3 kg and 0.1 kg are connected by a light inextensible string which passes over a smooth fixed pulley as shown.

The system is released from rest with the particles hanging vertically. Calculate:
(i) the acceleration of the system
(ii) the tension in the string
(iii) the force exerted by the string on the pulley when the particles are in motion.

The 0.3 kg particle will move down as it is heavier, while the 0.1 kg particle will move up.
The tension in the string acts away from each particle and is the same (you would call it T).
The weight of each particle is found using
$$F = ma$$
where a =the acceleration due to gravity = 10 m/s²

So the weight of the 0.3 kg particle is:

$$0.3 \times 10 = 3 \text{ N}$$

and the weight of the 0.1 kg particle is:

$$0.1 \times 10 = 1 \text{ N}$$

(i) To solve the question, we need to use these rules:
- Consider each particle separately.
- Use F = ma for each one.

First consider the 0.3 kg particle:

Using $\qquad$ F = ma

$$3 - T = 0.3a$$

Then consider the 0.1 kg particle:

$$T - 1 = 0.1a$$

You can then solve these simultaneous equations by adding them to get rid of the T's.

$$3 - T = 0.3a$$
$$+ \ \underline{T - 1 = 0.1a}$$
$$2 = 0.4a$$
$$a = 5 \text{ m/s}^2$$

(ii) You can find the tension by substituting the acceleration into either equation:

$$T - 1 = 0.1a$$
$$T - 1 = 0.5$$
$$T = 1.5 \text{ N}$$

(ii) The force exerted by the string on the pulley when the particles are in motion will be the sum of the tensions:

$$1.5 + 1.5 = 3 \text{ N}$$

Exercise 21D:

1. Two particles of mass 0.9kg and 0.4 kg are connected by a light inextensible string which passes over a smooth fixed pulley. The system is released from rest with the particles hanging vertically. Calculate:
   (i) the acceleration of the system
   (ii) the tension in the string
   (iii) the force exerted by the string on the pulley when the particles are in motion.

2. Two particles of mass 1.2 kg and 0.8 kg are connected by a light inextensible string which passes over a smooth fixed pulley. The system is released from rest with the particles hanging vertically. Calculate:
   (i) the acceleration of the system
   (ii) the tension in the string
   (iii) the force exerted by the string on the pulley when the particles are in motion.

3. Two particles of mass 0.3 kg and 0.2 kg are connected by a light inextensible string which passes over a smooth fixed pulley. The system is released from rest with the particles hanging

Exercise 21D...

vertically. Calculate:
(i) the acceleration of the system
(ii) the tension in the string
(iii) the force exerted by the string on the pulley when the particles are in motion.

4. Two particles of mass 0.7 kg and 0.5 kg are connected by a light inextensible string which passes over a smooth fixed pulley. The system is released from rest with the particles hanging vertically. Calculate:
   (i) the acceleration of the system
   (ii) the tension in the string
   (iii) the force exerted by the string on the pulley when the particles are in motion.

5. Two particles of mass 1.5 kg and 1 kg are connected by a light inextensible string which passes over a smooth fixed pulley. The system is released from rest with the particles hanging vertically. Calculate:
   (i) the acceleration of the system
   (ii) the tension in the string
   (iii) the force exerted by the string on the pulley when the particles are in motion.

6. Two particles of mass 1.1 kg and 0.5 kg are connected by a light inextensible string which passes over a smooth fixed pulley. The system is released from rest with the particles hanging vertically. Calculate:
   (i) the acceleration of the system
   (ii) the tension in the string
   (iii) the force exerted by the string on the pulley when the particles are in motion.

7. Two particles of mass 0.5 kg and 0.3 kg are connected by a light inextensible string which passes over a smooth fixed pulley. The system is released from rest with the particles hanging vertically. Calculate:
   (i) the acceleration of the system
   (ii) the tension in the string
   (iii) the force exerted by the string on the pulley when the particles are in motion.

8. Two particles of mass 70 kg and 30 kg are connected by a light inextensible string which passes over a smooth fixed pulley. The system is released from rest with the particles hanging vertically. Calculate:
   (i) the acceleration of the system
   (ii) the tension in the string
   (iii) the force exerted by the string on the pulley when the particles are in motion.

21.5 Pulleys – Harder Questions

Example:

Two particles A and B of mass 0.7 kg and 0.2 kg respectively are connected by a light inextensible string which passes over a smooth fixed pulley as shown. Particle A is 0.72m above the ground. The system is released from rest with the particles hanging vertically. Calculate:

(i) the acceleration of the system

(ii) the tension in the string

(iii) the force exerted by the string on the pulley when the particles are in motion.

When A hits the ground the string becomes slack and B continues to rise. Assuming that B does not reach the pulley, find:

(iv) the speed of the particles at the instant A hits the ground

(v) the additional distance B rises after A hits the ground

(vi) the time that elapses between A hitting the ground and the string becoming taut again.

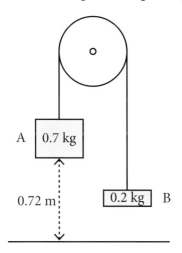

Parts (i) (ii) and (iii) are worked out as before.

The weight of the 0.7 kg particle is:

$$0.7 \times 10 = 7 \text{ N}$$

and the weight of the 0.2 kg particle is:

$$0.2 \times 10 = 1 \text{ N}$$

Let the tension in the string be T. Since A is heavier than B, it will move downwards, and B will move upwards.

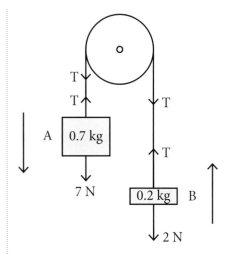

(i) First consider particle A:
Using $\qquad$ F = ma
$$7 - T = 0.7a$$
Then consider particle B:
Using $\qquad$ F = ma
$$T - 2 = 0.2a$$
Adding these quations together gives:
$$5 = 0.9a$$
$$a = \frac{5}{0.9}$$
$$a = 5.56 \text{ m/s}^2$$

(ii) Substitute a into either equation to get T:
$$T - 2 = 0.2a$$
$$T - 2 = 0.2 \times 5.56$$
$$T - 2 = 1.112$$
$$T = 3.11 \text{ N}$$

(iii) The force on pulley is the sum of the tensions:
$$= 2 \times 3.112$$
$$= 6.224 \text{ N}$$

(iv) The system is released from rest. So we know that
$$u = 0$$
$$a = 5.56$$
$$s = 0.72$$
We want to find v = ?
So we use $\qquad$ $v^2 = u^2 + 2as$
$$v^2 = 0 + 2(5.56)(0.72)$$
$$v^2 = 8.0064$$
$$v = \sqrt{8.0064}$$
$$v = 2.83 \text{ m/s}$$

(v) After A hits the ground the string becomes slack and B continues to rise. The acceleration slowing B down is the acceleration due to gravity which is –10 m/s². The initial speed of B after A hits the ground equals

the speed of the particles at the instant A hits the ground = 2.83 m/s. B will rise until it stops, ie until v = 0.
So we know   u = 2.83

          a = −10

          v = 0

We want to find s = ?
So we use $v^2 = u^2 + 2as$

          $0 = 2.83^2 − 2(10)s$

          20s = 8.009 kg

          s = 0.4 m

(vi) After B stops it will then begin to fall down again under gravity until it reaches the speed of 2.83 m/s. At this point the string will become taut again. You need to find the time taken until B stops and then double this (since it moved up, and then down again, after A hit the ground).
We know   u = 2.83

          a = −10

          v = 0

We want to find t = ?
So we use   v = u + at

          0 = 2.83 − 10t

          10t = 2.83

          t = 0.283 s

So the time that elapses between A hitting the ground and the string becoming taut again is:

          0.283 × 2 = 0.566 s

Exercise 21E:

1. Two particles A and B of mass 0.6 kg and 0.4 kg respectively are connected by a light inextensible string which passes over a smooth fixed pulley. Particle A is 1.2m above the ground. The system is released from rest with the particles hanging vertically. Calculate:
(i) the acceleration of the system
(ii) the tension in the string
(iii) the force exerted by the string on the pulley when the particles are in motion.
When A hits the ground the string becomes slack and B continues to rise. Assuming that B does not reach the pulley find:
(iv) the time it takes for A to hit the ground
(v) the speed of the particles at the instant A hits the ground
(vi) the additional distance B rises after A hits the ground
(vii) the time that elapses between A hitting the ground and the string becoming taut again.

Exercise 21E...

2. Two particles A and B of mass 2.6 kg and 1.4 kg respectively are connected by a light inextensible string which passes over a smooth fixed pulley. The system is released from rest with the particles hanging vertically. Calculate:
(i) the acceleration of the system
(ii) the tension in the string
(iii) the force exerted by the string on the pulley when the particles are in motion.
A hits the ground after 1.2 s. When A hits the ground the string becomes slack and B continues to rise. Assuming that B does not reach the pulley find:
(iv) the distance that A falls to the ground
(v) the speed of the particles at the instant A hits the ground
(vi) the additional distance B rises after A hits the ground
(vii) the time that elapses between A hitting the ground and the string becoming taut again.

3. Two particles A and B of mass 0.7 kg and 0.3 kg respectively are connected by a light inextensible string which passes over a smooth fixed pulley. The system is released from rest with the particles hanging vertically. Calculate:
(i) the acceleration of the system
(ii) the tension in the string
(iii) the force exerted by the string on the pulley when the particles are in motion.
A hits the ground after 1.5 s. When A hits the ground the string becomes slack and B continues to rise. Assuming that B does not reach the pulley find:
(iv) the distance that A falls to the ground
(v) the speed of the particles at the instant A hits the ground
(vi) the additional distance B rises after A hits the ground
(vii) the time that elapses between A hitting the ground and the string becoming taut again.

4. Two particles A and B of mass 3.6 kg and 1.4 kg respectively are connected by a light inextensible string which passes over a smooth fixed pulley. Particle A is 0.75m above the ground. The system is released from rest with the particles hanging vertically. Calculate:
(i) the acceleration of the system
(ii) the tension in the string
(iii) the force exerted by the string on the pulley

when the particles are in motion.
When A hits the ground the string becomes slack and B continues to rise. Assuming that B does not reach the pulley find:
(iv) the time it takes for A to hit the ground
(v) the speed of the particles at the instant A hits the ground
(vi) the additional distance B rises after A hits the ground
(vii) the time that elapses between A hitting the ground and the string becoming taut again.

5. Two particles A and B of mass 1.3 kg and 0.7 kg respectively are connected by a light inextensible string which passes over a smooth fixed pulley. The system is released from rest with the particles hanging vertically. Calculate:
(i) the acceleration of the system
(ii) the tension in the string
(iii) the force exerted by the string on the pulley when the particles are in motion.
A hits the ground after 0.8 s. When A hits the ground the string becomes slack and B continues to rise. Assuming that B does not reach the pulley find:
(iv) the distance that A falls to the ground
(v) the speed of the particles at the instant A hits the ground
(vi) the additional distance B rises after A hits the ground
(vii) the time that elapses between A hitting the ground and the string becoming taut again.

6. Two particles A and B of mass 3.7 kg and 1.3 kg respectively are connected by a light inextensible string which passes over a smooth fixed pulley. The system is released from rest with the particles hanging vertically. Calculate:
(i) the acceleration of the system
(ii) the tension in the string
(iii) the force exerted by the string on the pulley when the particles are in motion.
A hits the ground after 0.4 s. When A hits the ground the string becomes slack and B continues

to rise. Assuming that B does not reach the pulley find:
(iv) the distance that A falls to the ground
(v) the speed of the particles at the instant A hits the ground
(vi) the time that elapses between A hitting the ground and the string becoming taut again.

7. Two particles A and B of mass 1.5 kg and 0.5 kg respectively are connected by a light inextensible string which passes over a smooth fixed pulley. The system is released from rest with the particles hanging vertically. Calculate:
(i) the acceleration of the system
(ii) the tension in the string
(iii) the force exerted by the string on the pulley when the particles are in motion.
A hits the ground at 6 m/s. When A hits the ground the string becomes slack and B continues to rise. Assuming that B does not reach the pulley find:
(iv) the time it takes for A to hit the ground
(v) the distance that A falls to the ground
(vi) the time that elapses between A hitting the ground and the string becoming taut again.

8. Two particles A and B of mass 2.4 kg and 0.6 kg respectively are connected by a light inextensible string which passes over a smooth fixed pulley. The system is released from rest with the particles hanging vertically. Calculate:
(i) the acceleration of the system
(ii) the tension in the string
(iii) the force exerted by the string on the pulley when the particles are in motion.
A hits the ground at 3 m/s. When A hits the ground the string becomes slack and B continues to rise. Assuming that B does not reach the pulley find:
(iv) the time it takes for A to hit the ground
(v) the distance that A falls to the ground
(vi) the time that elapses between A hitting the ground and the string becoming taut again.

21.6 Pulleys Where One Body Is On A Horizontal Table

You can also be asked to deal with problems involving pulleys, where only one object hangs vertically and the other is on a horizontal surface. For example:

Two particles of mass 4 kg and 6 kg are connected by a light inextensible string which passes over a smooth fixed pulley. The 4 kg body lies on a smooth horizontal table and the 6 kg body hangs vertically as shown.

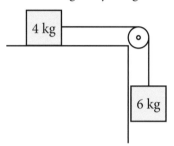

The system is released from rest. Calculate:
(i) the acceleration of the system
(ii) the tension in the string
(iii) the force exerted by the string on the pulley when the particles are in motion.

Because the table is smooth there is no friction acting on the 4 kg body. When the system is released from rest the 6 kg body will move downwards and the 4 kg body will move to the right. You need to mark on all the forces acting in the bodies.

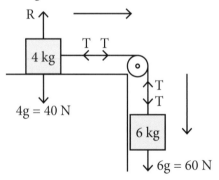

The rules for solving this problem are similar to before:
• Consider each particle separately.
• Use F = ma
(i) First, consider the 6 kg particle:
$$F = ma$$
$$60 - T = 6a$$
Next consider the 4 kg particle:
$$F = ma$$
$$T = 4a$$
Note that neither the reaction R nor the 40 N weight affect the horizontal motion as they are perpendicular to the line of motion.

You can then solve these simultaneous equations by adding them to get rid of the T's:

$$60 - T = 6a$$
$$+ \quad\quad T = 4a$$
$$60 = 10a$$
$$a = 6 \text{ m/s}^2$$

(ii) You can find the tension by substituting into either equation:
$$T = 4a$$
$$T = 4 \times 6$$
$$= 24 \text{ N}$$

(iii) The two tensions act on the pulley. Since these tensions are vectors you must add them nose-to-tail and then work out the resultant force.

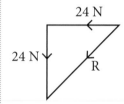

$$R^2 = 24^2 + 24^2$$
$$R^2 = 576 + 576$$
$$R^2 = 1152$$
$$R = \sqrt{1152}$$
$$R = 33.94....$$
So $\quad R = 33.9$ N to 3 significant figures

Exercise 21F:

1. Two particles of mass 7 kg and 8 kg are connected by a light inextensible string which passes over a smooth fixed pulley. The 7 kg body lies on a smooth horizontal table and the 8 kg body hangs vertically. The system is released from rest. Calculate:
(i) the acceleration of the system correct to 3 significant figures
(ii) the tension in the string correct to 3 significant figures
(iii) the force exerted by the string on the pulley when the particles are in motion.

2. Two particles of mass 2 kg and 5 kg are connected by a light inextensible string which passes over a smooth fixed pulley. The 2 kg body lies on a smooth horizontal table and the 5 kg body hangs vertically. The system is released from rest. Calculate:
(i) the acceleration of the system correct to 3 significant figures

(ii) the tension in the string correct to 3 significant figures

(iii) the force exerted by the string on the pulley when the particles are in motion.

3. Two particles of mass 3 kg and 8 kg are connected by a light inextensible string which passes over a smooth fixed pulley. The 3 kg body lies on a smooth horizontal table and the 8 kg body hangs vertically. The system is released from rest. Calculate:

(i) the acceleration of the system correct to 3 significant figures

(ii) the tension in the string correct to 3 significant figures

(iii) the force exerted by the string on the pulley when the particles are in motion.

4. Two particles of mass 0.4 kg and 0.7 kg are connected by a light inextensible string which passes over a smooth fixed pulley. The 0.4 kg body lies on a smooth horizontal table and the 0.7 kg body hangs vertically. The system is released from rest. Calculate:

(i) the acceleration of the system correct to 3 significant figures

(ii) the tension in the string correct to 3 significant figures

(iii) the force exerted by the string on the pulley when the particles are in motion.

5. Two particles of mass 2.4 kg and 3.2 kg are connected by a light inextensible string which passes over a smooth fixed pulley. The 2.4 kg body lies on a smooth horizontal table and the 3.2 kg body hangs vertically. The system is released from rest. Calculate:

(i) the acceleration of the system correct to 3 significant figures

(ii) the tension in the string correct to 3 significant figures

(iii) the force exerted by the string on the pulley when the particles are in motion.

6. Two particles of mass 16 kg and 22 kg are connected by a light inextensible string which passes over a smooth fixed pulley. The 16 kg body lies on a smooth horizontal table and the 22 kg body hangs vertically. The system is released from rest. Calculate:

(i) the acceleration of the system correct to 3 significant figures

(ii) the tension in the string correct to 3 significant figures

(iii) the force exerted by the string on the pulley when the particles are in motion.

7. Two particles of mass 4.5 kg and 6 kg are connected by a light inextensible string which passes over a smooth fixed pulley. The 4.5 kg body lies on a smooth horizontal table and the 6 kg body hangs vertically. The system is released from rest. Calculate:

(i) the acceleration of the system correct to 3 significant figures

(ii) the tension in the string correct to 3 significant figures

(iii) the force exerted by the string on the pulley when the particles are in motion.

8. Two particles of mass 0.8 kg and 1.2 kg are connected by a light inextensible string which passes over a smooth fixed pulley. The 0.8 kg body lies on a smooth horizontal table and the 1.2 kg body hangs vertically. The system is released from rest. Calculate:

(i) the acceleration of the system

(ii) the tension in the string

(iii) the force exerted by the string on the pulley when the particles are in motion.

9. Two particles of mass 7 kg and 9 kg are connected by a light inextensible string which passes over a smooth fixed pulley. The 7 kg body lies on a smooth horizontal table and the 9 kg body hangs vertically. The system is released from rest. Calculate:

(i) the acceleration of the system

(ii) the tension in the string

(iii) the force exerted by the string on the pulley when the particles are in motion.

10. Two particles of mass 48 kg and 60 kg are connected by a light inextensible string which passes over a smooth fixed pulley. The 48 kg body lies on a smooth horizontal table and the 60 kg body hangs vertically. The system is released from rest. Calculate:

(i) the acceleration of the system correct to 3 significant figures

(ii) the tension in the string correct to 4 significant figures

(iii) the force exerted by the string on the pulley when the particles are in motion.

21.7 Pulleys Where One Body Is On A Horizontal Table – Harder Questions

You can be asked more complex questions involving pulleys and horizontal surfaces, which often require you to take friction into account. For example:

Two particles A and B, of mass 0.8 kg and 1.4 kg respectively, are connected by a light inextensible string which passes over a smooth fixed pulley. Particle A lies on a rough horizontal table and particle B hangs vertically as shown exactly 0.75m above the ground.

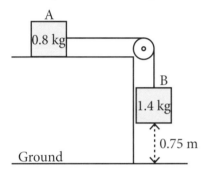

The coefficient of limiting friction is 0.46
The system is released from rest. Calculate:
(i) the acceleration of the system
(ii) the tension in the string
(iii) the velocity with which particle B strikes the ground
After particle B strikes the ground particle A continues to move. Assuming A does not reach the pulley find:
(iv) the time it takes A to stop
(v) the distance A travels before stopping.

Because the table is rough there is friction acting on the 4 kg body. When the system is released from rest the 1.4 kg body will move downwards and the 0.8 kg body will move to the right. You need to mark on all the forces acting in the bodies including the frictional force, F, acting on the 0.8 kg body.

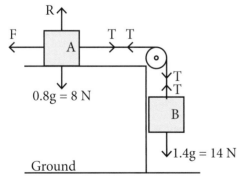

The rules for solving this type of problem are:
• Consider each particle separately.
• Work out the frictional force acting on the 0.8 kg body using $F = \mu R$
• Then use $F = ma$

(i) First consider the 1.4 kg particle:
$$F = ma$$
$$14 - T = 1.4a$$
Then consider the 0.8 kg particle:
Resolve vertically $R = 8$ N
Frictional force $F = \mu R$
$$F = 0.46 \times 8$$
$$F = 3.68 \text{ N}$$
You can then apply Newton's Law
$$F = ma$$
$$T - 3.68 = 0.8a$$
You can then solve these simultaneous equations by adding them to get rid of the T's:
$$14 - T = 1.4a$$
$$+ \ \underline{T - 3.68 = 0.8a}$$
$$10.32 = 2.2a$$
$$a = \frac{10.32}{2.2}$$
$$a = 4.690\ldots.. \text{ m/s}^2$$
$$a = 4.69 \text{ m/s}^2 \text{ to 3 significant figures.}$$

(ii) You can find the tension by substituting into either equation.
$$T - 3.68 = 0.8a$$
$$T - 3.68 = 0.8 \times 4.69$$
$$T - 3.68 = 3.752$$
$$T = 3.752 + 3.68$$
$$T = 7.43 \text{ N to 3 significant figures.}$$

(iii) You need to use the appropriate equation of motion
We know $\quad u = 0$ (ie, system is released from rest)
$$a = 4.69$$
$$s = 0.75$$
We want to find $v = ?$
So we use $\quad v^2 = u^2 + 2as$
$$v^2 = 0 + 2 \times 4.69 \times 0.75$$
$$v^2 = 7.035$$
$$v = \sqrt{7.035}$$
$$v = 2.652\ldots.$$
$$v = 2.65 \text{ m/s to 3 significant figures.}$$

(iv) At the moment the 1.4 kg body hits the ground, the 0.8 kg body is moving at 2.65 m/s. When it hits the ground the string will become slack and so there will be no tension pulling the 0.8 kg body. The only horizontal force acting on the 0.8 kg body will be the frictional force, as shown below, which will slow it down.

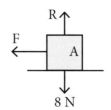

This gives

$$F = ma$$
$$-3.68 = 0.8a$$

So

$$a = \frac{-3.68}{0.8}$$
$$a = -4.6 \text{ m/s}^2$$

Thus we know the following:

$$u = 2.65$$
$$a = -4.6$$
$$v = 0 \text{ (when it stops the velocity is 0)}$$

We want to find $t = ?$

So we use

$$v = u + at$$
$$0 = 2.65 - 4.6t$$
$$4.6t = 2.65$$
$$t = \frac{2.65}{4.6}$$
$$= 0.5760....$$
$$t = 0.576 \text{ s to 3 significant figures}$$

(v) To find the distance travelled, we know:

$$u = 2.65$$
$$a = -4.6$$
$$v = 0$$

We want to find $s = ?$

So we use

$$v^2 = u^2 + 2as$$
$$0 = 2.65^2 - 2 \times 4.6 \times s$$
$$0 = 7.0225 - 9.2s$$
$$9.2s = 7.0225$$
$$s = \frac{7.0225}{9.2}$$
$$s = 0.7633....$$
$$s = 0.763 \text{ m to 3 significant figures}$$

Exercise 21G:

1. Two particles A and B, of mass 2 kg and 5 kg respectively, are connected by a light inextensible string which passes over a smooth fixed pulley. Particle A lies on a rough horizontal table and particle B hangs vertically. The coefficient of limiting friction is 0.8. The system is released from rest. Calculate:
(i) the acceleration of the system
(ii) the tension in the string
B hits the ground after 0.48 s. Calculate:
(iii) the velocity with which particle B strikes the ground
After particle B strikes the ground particle A continues to move. Assuming A does not reach the pulley find:
(iv) the time it takes A to stop
(v) the distance A travels before stopping.

Exercise 21G...

2. Two particles A and B, of mass 0.4 kg and 0.7 kg respectively, are connected by a light inextensible string which passes over a smooth fixed pulley. Particle A lies on a rough horizontal table and particle B hangs vertically exactly 0.64m above the ground. The coefficient of limiting friction is $\frac{3}{8}$. The system is released from rest. Calculate:
(i) the acceleration of the system
(ii) the tension in the string
(iii) the velocity with which particle B strikes the ground.
After particle B strikes the ground particle A continues to move. Assuming A does not reach the pulley find:
(iv) the time it takes A to stop
(v) the distance A travels before stopping.

3. Two particles A and B, of mass 16 kg and 24 kg respectively, are connected by a light inextensible string which passes over a smooth fixed pulley. Particle A lies on a rough horizontal table and particle B hangs vertically. The coefficient of limiting friction is 0.6
The system is released from rest. Calculate:
(i) the acceleration of the system
(ii) the tension in the string
B hits the ground after 0.42 s. Calculate
(iii) the velocity with which particle B strikes the ground.
After particle B strikes the ground particle A continues to move. Assuming A does not reach the pulley find:
(iv) the time it takes A to stop
(v) the distance A travels before stopping.

4. Two particles A and B, of mass 1.3 kg and 1.8 kg respectively, are connected by a light inextensible string which passes over a smooth fixed pulley. Particle A lies on a rough horizontal table and particle B hangs vertically exactly 1.2m above the ground. The coefficient of limiting friction is $\frac{2}{5}$. The system is released from rest. Calculate:
(i) the acceleration of the system
(ii) the tension in the string
(iii) the velocity with which particle B strikes the ground.
After particle B strikes the ground particle A continues to move. Assuming A does not reach the pulley find:

(iv) the time it takes A to stop

(v) the distance A travels before stopping.

5. Two particles A and B, of mass 45 kg and 60 kg respectively, are connected by a light inextensible string which passes over a smooth fixed pulley. Particle A lies on a rough horizontal table and particle B hangs vertically. The coefficient of limiting friction is 0.3. The system is released from rest. Calculate:

(i) the acceleration of the system

(ii) the tension in the string

B hits the ground after 0.24 s. Calculate:

(iii) the velocity with which particle B strikes the ground.

After particle B strikes the ground particle A continues to move. Assuming A does not reach the pulley find:

(iv) the time it takes A to stop

(v) the distance A travels before stopping.

6. Two particles A and B, of mass 18 kg and 21 kg respectively, are connected by a light inextensible string which passes over a smooth fixed pulley. Particle A lies on a rough horizontal table and particle B hangs vertically exactly 0.45m above the ground.

The coefficient of limiting friction is $\frac{3}{10}$. The system is released from rest. Calculate:

(i) the acceleration of the system

(ii) the tension in the string

(iii) the velocity with which particle B strikes the ground.

After particle B strikes the ground particle A continues to move. Assuming A does not reach the pulley find:

(iv) the time it takes A to stop

(v) the distance A travels before stopping.

7. Two particles A and B, of mass 0.64 kg and 0.72 kg respectively, are connected by a light inextensible string which passes over a smooth fixed pulley. Particle A lies on a rough horizontal table and particle B hangs vertically. The coefficient of limiting friction is 0.42 The system is released from rest. Calculate:

(i) the acceleration of the system

(ii) the tension in the string

B hits the ground after 0.76 s. Calculate:

(iii) the velocity with which particle B strikes the ground.

After particle B strikes the ground particle A continues to move. Assuming A does not reach the pulley find:

(iv) the time it takes A to stop

(v) the distance A travels before stopping.

8. Two particles A and B, of mass 9 kg and 12 kg respectively, are connected by a light inextensible string which passes over a smooth fixed pulley. Particle A lies on a rough horizontal table and particle B hangs vertically exactly 0.85m above the ground. The coefficient of limiting friction is 0.36. The system is released from rest. Calculate:

(i) the acceleration of the system

(ii) the tension in the string

(iii) the velocity with which particle B strikes the ground.

After particle B strikes the ground particle A continues to move. Assuming A does not reach the pulley find:

(iv) the time it takes A to stop

(v) the distance A travels before stopping.

9. Two particles A and B, of mass 36 kg and 42 kg respectively, are connected by a light inextensible string which passes over a smooth fixed pulley. Particle A lies on a rough horizontal table and particle B hangs vertically. The coefficient of limiting friction is $\frac{3}{7}$. The system is released from rest. Calculate:

(i) the acceleration of the system

(ii) the tension in the string

B hits the ground after 0.76 s. Calculate:

(iii) the velocity with which particle B strikes the ground.

After particle B strikes the ground particle A continues to move. Assuming A does not reach the pulley find:

(iv) the time it takes A to stop

(v) the distance A travels before stopping.

10. Two particles A and B, of mass 2.4 kg and 2.9 kg respectively, are connected by a light inextensible string which passes over a smooth fixed pulley. Particle A lies on a rough horizontal table and particle B hangs vertically exactly 1.16m above the ground. The coefficient of limiting friction is 0.72. The system is released from rest. Calculate:

(i) the acceleration of the system

(ii) the tension in the string

(iii) the velocity with which particle B strikes the ground.

After particle B strikes the ground particle A continues to move. Assuming A does not reach the pulley find:

(iv) the time it takes A to stop

(v) the distance A travels before stopping.

CHAPTER 22: MOMENTS

22.1 The Principle of Moments

The moment of a force measures the **turning effect** of the force. The moment is proportional to both the force and the perpendicular distance from the turning point. It can be calculated as follows:

Moment = Force × perpendicular distance

The **Principle of Moments** states that:
When a system is in equilibrium then the sum of all the clockwise moments about any point is equal to the sum of all the anticlockwise moments about this point.

22.2 Finding Two Unknown Forces

The rules when finding two unknown forces are:
- Take moments about the point where one of these forces acts.
- Equate the vertical forces.

For example:
AB is a uniform rod of length 8 m and mass 4.8 kg. It rests on two supports, one at C where AC = 1.6 m and the other at D where DB = 2 m. Find the reactions at C and D.

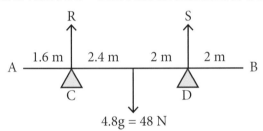

AB is a uniform rod and so the weight acts at the centre of AB. Therefore the weight = 4.8g = 48 N.
Taking moments about C gives:
$$S \times 4.4 = 48 \times 2.4$$
$$4.4S = 115.2$$
$$S = \frac{115.2}{4.4}$$
$$S = 26.2 \text{ N}$$
Equating forces gives:
$$R + S = 48$$
$$R + 26.2 = 48$$
$$R = 48 - 26.2$$
$$R = 21.8 \text{ N}$$

Exercise 22A:

1. AB is a uniform rod of length 12 m and mass 8 kg. It rests on two supports, one at A and the other at C where CB = 2 m. Find the reactions at A and C.

2. AB is a uniform rod of length 10 m and mass 5 kg It rests on two supports, one at C where AC = 2 m and the other at D where DB = 2.5 m. Find the reactions at C and D.

3. AB is a uniform rod of length 18 m and mass 6 kg. It rests on two supports, one at C where AC = 2 m and the other at B. Find the reactions at C and B.

4. AB is a uniform rod of length 24 m and mass 12 kg It rests on two supports, one at C where AC = 7 m and the other at D where DB = 9 m. Find the reactions at C and D.

5. AB is a uniform rod of length 16 m and mass 5.6 kg It rests on two supports, one at A and the other at C where CB = 3.5 m. Find the reactions at A and C.

6. AB is a uniform rod of length 10 m and mass 2.6 kg It rests on two supports, one at C where AC = 2.4 m and the other at D where DB = 3.3 m. Find the reactions at C and D.

7. AB is a uniform rod of length 12 m and mass 0.8 kg It rests on two supports, one at C where AC = 1.2 m and the other at D where DB = 2.3 m. Find the reactions at C and D.

8. AB is a uniform rod of length 14 m and mass 3.6 kg It rests on two supports, one at C where AC = 1.5 m and the other at D where DB = 3 m. Find the reactions at C and D.

Some questions feature rods suspended from strings, rather than resting on supports. They can be answered in the same way. For example:

A uniform rod AB of length 18 m and mass 7.8 kg is held horizontally in equilibrium by two strings, one at C where AC = 2 m and the other at D where DB = 6 m. Find the tensions in the two strings.

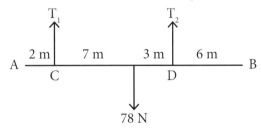

Taking moments about C gives:
$$T_2 \times 10 = 78 \times 7$$
$$10T_2 = 546$$
$$T_2 = 54.6 \text{ N}$$

Equating forces gives:
$$T_1 + T_2 = 78$$
$$T_1 + 54.6 = 78$$
$$T_1 = 78 - 54.6$$
$$T_1 = 23.4N$$

Exercise 22B:

1. A uniform rod AB of length 16 m and mass 5 kg is held horizontally in equilibrium by two strings, one at A and the other at C where AC = 14 m. Find the tensions in the two strings.

2. A uniform rod AB of length 28 m and mass 8 kg is held horizontally in equilibrium by two strings, one at C where AC = 2 m and the other at D where DB = 9 m. Find the tensions in the two strings.

3. A uniform rod AB of length 10 m and mass 2.4 kg is held horizontally in equilibrium by two strings, one at C where AC = 2 m and the other at B. Find the tensions in the two strings.

4. A uniform rod AB of length 12 m and mass 5.6 kg is held horizontally in equilibrium by two strings, one at A and the other at C where AC = 10 m. Find the tensions in the two strings.

5. A uniform rod AB of length 16 m and mass 0.9 kg is held horizontally in equilibrium by two strings, one at C where AC = 3 m and the other at D where DB = 6 m. Find the tensions in the two strings.1.5

6. A uniform rod AB of length 8 m and mass 1.6 kg is held horizontally in equilibrium by two strings, one at C where AC = 1.5 m and the other at B. Find the tensions in the two strings.

7. A uniform rod AB of length 24 m and mass 3 kg is held horizontally in equilibrium by two strings, one at A and the other at C where AC = 14.5 m. Find the tensions in the two strings.

8. A uniform rod AB of length 8 m and mass 24 kg is held horizontally in equilibrium by two strings, one at C where AC = 2.4 m and the other at D where DB = 2.2 m. Find the tensions in the two strings.

9. A uniform rod AB of length 32 m and mass 0.4 kg is held horizontally in equilibrium by two strings, one at C where AC = 3 m and the other at D where DB = 12 m. Find the tensions in the two strings.

10. A uniform rod AB of length 30 m and mass 3.6 kg is held horizontally in equilibrium by two strings, one at C where AC = 7 m and the other at B. Find the tensions in the two strings.

22.3 Finding An Unknown Force And A Distance

If a distance is unknown, you can use these rules:
• Equate the vertical forces.
• Take moments about the point where one of these forces acts.
For example:

AB is a uniform rod of length 16 m and mass 5.5 kg
It rests on two supports, one at C where AC = 1.6 m and the other at D where DB = 0.8 m. Find how far from C a mass of 3.2 kg should be placed so that the reaction at C will be twice the reaction at B.

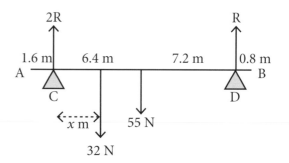

Equating forces gives:
$$2R + R = 32 + 55$$
$$3R = 87$$
$$R = 29$$

Taking moments about C gives:
$$32x + 55 \times 6.4 = 29 \times 13.6$$
$$32x + 352 = 394.4$$
$$32x = 394.4 - 352$$
$$32x = 42.4$$
$$x = 1.325$$
So the answer is: 1.325 m from C

Exercise 22C:

1. AB is a uniform rod of length 13 m and mass 2.4 kg It rests on two supports, one at C where AC = 2 m and the other at D where DB = 3.2 m. Find how far from C a mass of 1.8 kg should be placed so that the reaction at D will be twice the reaction at C.

2. AB is a uniform rod of length 20 m and mass 1.8 kg It rests on two supports, one at C where AC = 3 m and the other at D where DB = 6 m. Find how far from C a mass of 1.2 kg should be placed so that the reaction at D will be twice the reaction at C.

3. AB is a uniform rod of length 16 m and mass 3.8 kg It rests on two supports, one at C where AC = 3.5 m

Exercise 22C...

and the other at D where DB = 5.6 m. Find how far from C a mass of 3.4 kg should be placed so that the reaction at D will be twice the reaction at C.

4. AB is a uniform rod of length 18 m and mass 4 kg It rests on two supports, one at C where AC = 2 m and the other at D where DB = 6 m. Find how far from C a mass of 1.4 kg should be placed so that the reaction at D will be three times the reaction at C.

5. AB is a uniform rod of length 26 m and mass 2.8 kg It rests on two supports, one at C where AC = 4 m and the other at D where DB = 11 m. Find how far from C a mass of 1.7 kg should be placed so that the reaction at D will be twice the reaction at C.

6. AB is a uniform rod of length 8 m and mass 4.6 kg It rests on two supports, one at C where AC = 1.7 m and the other at D where DB = 2.8 m. Find how far from C a mass of 3.8 kg should be placed so that the reaction at D will be twice the reaction at C.

22.4 A Rod Just Starting To Tilt

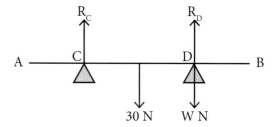

In the diagram above AB is a uniform rod of mass 3 kg It rests on two supports, one at C and the other at D. A weight W N is placed at D. As W is moved towards B the reactions change, with the reaction at D increasing as W approaches it and the reaction at C decreasing. There might come a time when the rod will **just start to tilt** about the reaction at D. At this moment the reaction at C will be 0 as shown below.

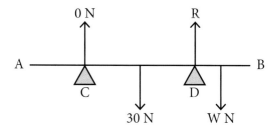

These types of problem can often be solved by taking moments around the pivot point. For example:

1. AB is a uniform rod of length 8 m and mass 5.6 kg It rests on two supports, one at C where AC = 1.6 m and the other at D where DB = 1.3 m. Find the largest mass that can be placed at B without the rod tilting.

Call the mass M kg. The reaction at C will be 0.

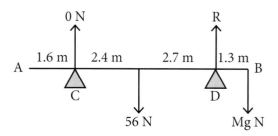

You should take moments about D so that you don't need to work out the reaction R.
Taking moments about D gives:
$$10M \times 1.3 = 56 \times 2.7$$
$$13M = 151.2$$
$$M = 11.6 \text{ kg}$$

2. AB is a uniform rod of length 14 m and mass 2.4 kg It rests on two supports, one at C where AC = 3 m and the other at D where DB = 3.2 m. Find how far from B a mass of 3.6 kg should be placed so that the rod will just start to tilt.

Call the distance from the 3.6 kg mass to D x metres. The reaction at C will be 0.

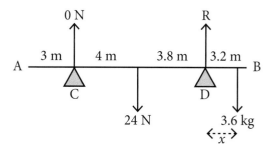

You should take moments about D so that you don't need to work out the reaction R.
Taking moments about D gives:
$$36x = 24 \times 3.8$$
$$36x = 91.2$$
$$x = 2.53 \text{ m}$$
However, this is the distance of the mass from D.
The distance from B will then be
$$3.2 - 2.53 = 0.67 \text{ m}$$

Exercise 22D:

1. AB is a uniform rod of length 12 m and mass 4.2 kg. It rests on two supports, one at C where AC = 2.5 m and the other at D where DB = 1.2 m. Find the largest mass that can be placed at B without the rod tilting.

2. AB is a uniform rod of length 18 m and mass 3.5 kg. It rests on two supports, one at C where AC = 2 m and the other at D where DB = 3 m. Find how far from B a mass of 8.4 kg should be placed so that the rod will just start to tilt.

3. AB is a uniform rod of length 6 m and mass 4.6 kg. It rests on two supports, one at C where AC = 1.2 m and the other at D where DB = 0.8 m. Find the largest mass that can be placed at B without the rod tilting.

4. AB is a uniform rod of length 24 m and mass 3.2 kg. It rests on two supports, one at C where AC = 4 m and the other at D where DB = 9 m. Find how far from B a mass of 2.6 kg should be placed so that the rod will just start to tilt.

5. AB is a uniform rod of length 10 m and mass 4.9 kg. It rests on two supports, one at C where AC = 2 m and the other at D where DB = 3.6 m. Find the largest mass that can be placed at B without the rod tilting.

6. AB is a uniform rod of length 28 m and mass 5.2 kg. It rests on two supports, one at C where AC = 4 m and the other at D where DB = 10.5 m. Find how far from B a mass of 4.3 kg should be placed so that the rod will just start to tilt.

7. A uniform rod AB of mass 12 kg and length 6 cm rests on two supports at C and D where AC = 2 cm and DB = 1.5 cm. Find:
   (a) the reactions at each support.
   (b) the least mass that should be placed at A to cause the rod to tilt.

CHAPTER 23: UNDERSTANDING AND USING STATISTICAL TERMINOLOGY

23.1 Statistical Definitions

A **population** is everything or everybody being included in an investigation or survey, for example the number of people in Northern Ireland or the number of football results in every league in Northern Ireland last year.

A **sample** is a smaller subset of the population, for example 1% of the number of people in Northern Ireland or the number of football results in every league in Northern Ireland in October last year. The larger the sample then the more accurate will be any conclusions drawn from the sample. There are different reasons for sampling:

1. It might be physically impossible to test the whole population, for example the number of people in Northern Ireland is continually changing with births, deaths etc.

2. Testing the population is possible, but would be too costly. For example, an airline company wants to ask its customers how helpful the staff were. They could ask everyone using their service but this would be too costly in terms of time, personnel, analysing the data collected.

3. Testing the population destroys the product. For example, a light bulb manufacturer could test how long each light bulb will last but in doing so there would be no product left to sell.

A **variable** is a property or a descriptive term that can take or have different values, for example the hair colours of the pupils in a class, the ages of the pupils in a class, the names of the pupils in a class, or the heights of the pupils in a class. A **discrete variable** is a variable that can only take certain values, such as the hair colours of the pupils in a class or the names of the pupils in a class. A **continuous variable** is a variable that can take any value within the range of measurements appropriate to that variable, for example the ages of the pupils in a class or the heights of the pupils in a class. All measurements are continuous variables.

Exercise 23A:

Suggest possible samples for the following surveys and populations:

	Survey	Population
1.	How much physical activity do young people aged 11 to 16 do on average per week.	The students in a school for 11 to 16 year olds.
2.	The average number of people in a town who own a mobile phone.	All the people living in the town.
3.	The average number of crisps in a standard bag of crisps.	All the standard bags of crisps in a shop.
4.	The average number of trains throughout Northern Ireland that are late.	All the trains throughout Northern Ireland.
5.	The approximate number of senior citizens in Northern Ireland.	The population of Northern Ireland.

Look at the variables below and say which ones are (i) continuous (ii) discrete.

6. The marks of students in an exam.

7. The weights of boxes in a store room.

8. The age of people going to church.

9. The number of different channels on television.

10. The volume of different bottles of juice.

11. Give 3 different examples of a
(i) continuous (ii) discrete variable.

12. Give an example of when it might be physically impossible to test the whole population.

13. Give an example of when testing the population is possible but would be too costly.

14. Give an example of when testing the population would destroy the product.

15. Say whether the following are continuous or discrete variables:

(a) goals scored in hockey matches
(b) the lengths of hockey pitches
(c) times that hockey matches last
(d) number of spectators at hockey matches.

16. Say whether the following are continuous or discrete variables:

(a) number of songs on a CD
(b) length of the songs on a CD
(c) surface area of CDs
(d) weight of CDs
(e) number of people involved in making a CD.

17. Say whether the following are discrete or continuous variables:

(i) weight of pupils in Year 12
(ii) number of Year 12 pupils absent
(iii) age of each Year 12 pupil.

18. A company makes dishwashers. They want to test how long a dishwasher will last before it breaks down. Write down why it is not a good idea to test the population.

19. A Maths teacher wants to test the hypothesis 'Pupils enjoy Maths'. Write down:

(i) what is wrong with the following samples
(ii) how she could improve each sample.

Sample 1: She chooses 10 pupils from the Further Maths class.

Sample 2: She chooses 30 boys, one from each class in school.

Sample 3: She chooses 6 pupils from each Year 11 and Year 12 class in school.

23.2 Grouped Distributions

You would normally group data when the range of data is very large. For example, the data below are the marks of pupils in a test:

13 17 19 21 22 24 25 25 28 31 32 34 36
40 42 43 45 47 48 49 53 54 57 59 61 64
65 66 69 72 73 75 78 81 84 85 86 90 93

These can be grouped in the frequency table below:

Group	Frequency
10–19	3
20–29	6
30–39	4
40–49	7
50–59	4
60–69	5
70–79	4
80–89	4
90–99	2

Each group is called a **class**. Each class has two **class limits**. The **lower** class limit is the smallest number included in the class. The **upper** class limit is the biggest number included in the class.
For example, in the class 70–79 above:
• the lower class limit is 70
• the upper class limit is 79

The **lower class boundary** is the number halfway between the upper class limit of the previous class and the lower class limit of the current class. The **upper class boundary** is the number halfway between the upper class limit of the current class and the lower class limit of the next class.

For example in the class 70–79 above:
• the lower class boundary is halfway between 69 and 70 and so it is 69.5
• the upper class boundary is halfway between 79 and 80 and so it is 79.5

The **class mark** is the mark taken to represent the class. It is the midvalue between the limits of the class. You can work out the midvalue by adding the two class limits and dividing by 2.
For example, in the class 70–79 above the class mark is
$$(70 + 79) \div 2$$
$$= 149 \div 2$$
$$= 74.5$$

Age Distributions. Note that in age groupings age is given in completed years, eg 5–9 years means $5 \leq age < 10$ years.

23.3 Statistical Averages

A statistical average can mean the mean, median or mode.
The **mean** is found by:
• Adding up all the figures.
• Dividing by how many figures there are.

The **median** is the middle figure when the figures are put in order (if there is an odd number of figures).
For example, the median of 2, 4, 7, 8, 9 is 7.
If there are an even number of figures, the median is half way between the two figures in the middle when the figures are put in order.

For example, the median of 2, 4, 7, 8, 9, 12 is the value half way between 7 and 8 which is 7.5.

The **mode** is the figure that occurs the most often in the distribution.

For example, the mode of 2, 4, 7, 8, 8, 9 is 8.
You are often required to work with statistical averages.
For example:

1. The table below shows the weights of different objects

Weight, kg	Frequency
1–6	7
7–12	14
13–18	22
19–24	36
25–30	54
31–36	3

(a) What is the lower class limit of the modal class?
(b) What is the upper class boundary of the median class?
(c) What is the class mark of the second class?

(a) The modal class is the class with the highest frequency which is 25–30. So the lower class limit is **25**.
(b) The median class is the class which includes the middle number. The total frequency is 136.
$136 \div 2 = 68$ and so the median class is the class which includes the 68th number. This is the class **19–24**.
So the upper class boundary is halfway between 24 and 25 which is 24.5.
(c) Find the class mark by adding the two class limits and dividing by 2. The class mark is halfway between 7 and 12:
$$7 + 12 = 19$$
$$19 \div 2 = \mathbf{9.5}$$

....

2. The table below shows the heights of different objects. There are no boundaries between the classes.

Height h, cm	Frequency
$1 < h \leq 10$	21
$10 < h \leq 19$	32
$19 < h \leq 28$	26
$28 < h \leq 37$	30
$37 < h \leq 46$	14
$46 < h \leq 55$	7

(a) What is the lower class limit of the modal class?
(b) What is the upper class limit of the median class?
(c) What is the class mark of the final class?

(a) The modal class is the class with the highest frequency which is $10 < h \leq 19$. So the lower class limit is **10**.
(b) The median class is the class which includes the middle number. The total frequency is 130.
$130 \div 2 = 65$ and so the median class is the class which includes the 65th number which is $19 < h \leq 28$.
So the upper class limit is **28**.

(c) The class mark is halfway between 46 and 55.
$$46 + 55 = 101$$
$$101 \div 2 = \mathbf{50.5}$$

Exercise 23B:

1. The table below shows the weights of different objects.

Weight, kg	Frequency
1–6	13
7–12	31
13–18	29
19–24	33
25–30	17
31–36	5

(a) What is the lower class limit of the modal class?
(b) What is the upper class boundary of the median class?
(c) What is the class mark of the second class?

2. The table below shows the heights of different objects.

Height h, cm	Frequency
$1 < h \leq 10$	34
$10 < h \leq 19$	32
$19 < h \leq 28$	24
$28 < h \leq 37$	12
$37 < h \leq 46$	4
$46 < h \leq 55$	2

(a) What is the lower class limit of the modal class?
(b) What is the upper class limit of the median class?
(c) What is the class mark of the fifth class?

3. The table below shows the lengths of different objects.

Length, cm	Frequency
14–19	4
20–25	18
26–31	27
32–38	13
39–44	37
45–51	12

(a) What is the upper class limit of the modal class?
(b) What is the lower class boundary of the median class?
(c) What is the class mark of the second class?

4. The table below shows the volumes of different objects.

Volume V, litres	Frequency
4 < V ≤ 16	9
16 < V ≤ 28	17
28 < V ≤ 40	21
40 < V ≤ 52	32
52 < V ≤ 64	11
64 < V ≤ 76	35

(a) What is the upper class limit of the modal class?
(b) What is the lower class limit of the median class?
(c) What is the class mark of the final class?

5. The table below shows the areas of different objects.

Area cm²	Frequency
1–3	27
4–6	39
7–9	24
10–12	32
13–15	12
16–18	6

(a) What is the lower class limit of the modal class?
(b) What is the upper class boundary of the median class?
(c) What is the class mark of the second class?

6. The table below shows the widths of different objects.

Width w, m	Frequency
1 < w ≤ 7	13
7 < w ≤ 13	42
13 < w ≤ 19	52
19 < w ≤ 25	28
25 < w ≤ 31	33
31 < w ≤ 37	17

(a) What is the lower class limit of the modal class?
(b) What is the upper class limit of the median class?
(c) What is the class mark of the final class?

23.4 Histograms

A histogram is where the data is represented by the area of each column. You should always plot the boundaries or limits as appropriate on the **horizontal** axis. You should always plot the frequency density on the **vertical** axis.

The **frequency density** is worked out as follows:

Frequency Density (or FD) = $\dfrac{\text{frequency}}{\text{class width}}$

The **class width** is worked out as follows:
Class width
  = upper class boundary – lower class boundary
or
  = upper class limit – lower class limit
as appropriate.

For example:

1. The table below shows the heights of different objects.

Height h, cm	Frequency
1 < h ≤ 4	42
4 < h ≤ 10	66
10 < h ≤ 15	42
15 < h ≤ 35	252
35 < h ≤ 37	17

Show this data on a histogram.

As there are no boundaries you use the following rules:
• Work out the class width using
Class width = upper class limit – lower class limit

• Work out the frequency density using $\dfrac{\text{frequency}}{\text{class width}}$.

• Plot the frequency density on the vertical axis.
• Plot the limits on the horizontal axis.

So we have:

Height h, cm	Class Width	Frequency (f)	Frequency Density (FD)
1 < h ≤ 4	4 – 1 = 3	42	42 ÷ 3 = 14
4 < h ≤ 10	10 – 4 = 6	66	66 ÷ 6 = 11
10 < h ≤ 15	15 – 10 = 5	42	42 ÷ 5 = 8.4
15 < h ≤ 35	35 – 15 = 20	252	252 ÷ 20 = 12.6
35 < h ≤ 37	37 – 35 = 2	17	17 ÷ 2 = 8.5

The histogram is shown below.

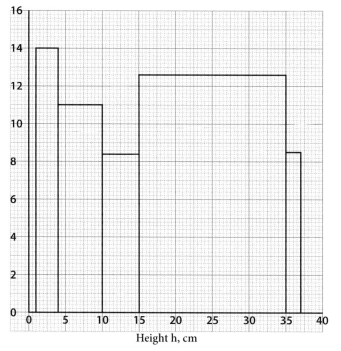

Height h, cm

2. The table below shows the lengths of different objects.

Length, cm	Frequency
1–5	18
6–7	11
8–17	24
18–22	22
23–32	18

Show this data on a histogram.

You use the following rules:
- Work out the class width using Class width
  = upper class boundary – lower class boundary
- Work out the frequency density using $\frac{\text{frequency}}{\text{class width}}$.
- Plot the frequency density on the vertical axis.
- Plot the limits on the horizontal axis.

So we have:

Length, cm	Class Width	Frequency (f)	Frequency Density (FD)
1 – 5	5.5 – 0.5 = 5	18	18 ÷ 5 = 3.6
6 – 7	7.5 – 5.5 = 2	11	11 ÷ 2 = 5.5
8 – 17	17.5 – 7.5 = 10	24	24 ÷ 10 = 2.4
18 – 22	22.5 – 17.5 = 5	22	22 ÷ 5 = 4.4
23 – 32	32.5 – 22.5 = 10	18	18 ÷ 10 = 1.8

The histogram is shown below.

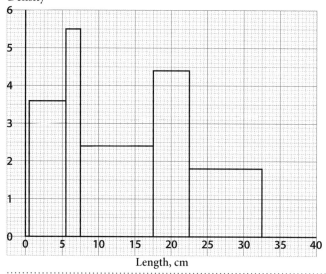

Length, cm

3. The table below shows the ages of different people.

Age, years	Frequency
10–49	18
50–59	11
60–84	24

Show this data on a histogram.

In age groupings, age is given in completed years:
So 10–49 years means 10 ≤ age < 50 years.
So 50–59 years means 50 ≤ age < 60 years.
So 60–84 years means 60 ≤ age < 85 years.

Thus we have:

Age, years	Class Width	Frequency	Frequency Density
10 ≤ age < 50	50 – 10 = 40	96	96 ÷ 40 = 2.4
50 ≤ age < 60	60 – 50 = 10	12	12 ÷ 10 = 1.2
60 ≤ age < 85	85 – 60 = 25	55	55 ÷ 25 = 2.2

The histogram is shown below.

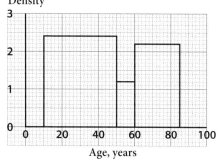

Age, years

Exercise 23C: *Draw histograms to show the following data:*

1.

Length, cm	Frequency
1–5	21
6–15	96
16–17	3
18–37	108

2.

Weight, kg	Frequency
1–5	710
6–15	890
16–17	262
18–22	240
23–37	2940

3.

Height h, cm	Frequency
$0 < h \le 5$	170
$5 < h \le 15$	640
$15 < h \le 17$	184
$17 < h \le 27$	320

4.

Volume, litres	Frequency
1–10	84
11–15	18
16–17	9
18–32	126
33–37	14

5.

Height h, cm	Frequency
$136 < h \le 138$	24
$138 < h \le 142$	52
$142 < h \le 147$	43
$147 < h \le 152$	46

6.

Diameter d, cm	Frequency
$16 < d \le 36$	68
$36 < d \le 41$	11
$41 < d \le 51$	46
$51 < d \le 60$	18

7.

Area,v cm²	Frequency
1 – 5	420
6 – 7	92
8 – 17	720
18 – 22	175

8.

Height h, cm	Frequency
$1 < h \le 6$	32
$6 < h \le 11$	19
$11 < h \le 15$	34
$15 < h \le 25$	27

9. The table below shows the ages of different people.

Age, years	Frequency
5–14	260
15–34	160
35–49	210

Show this data on a histogram.

10. The table below shows the ages of different people.

Age, years	Frequency
9–13	7
14–15	5
16–20	4

Show this data on a histogram.

23.5 Further Histograms

Sometimes you are given the histogram, and are asked to work out data from it. For example:

1. The histogram below shows the lengths of different objects in cm.

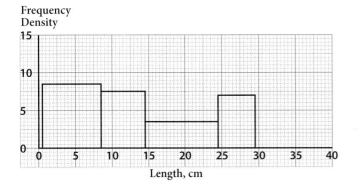

Frequency Density

Length, cm

(a) Complete the frequency table:

Length, cm	Frequency
1 – 8	

(b) What are the limits of the modal group?
(c) What are the boundaries of the median group?

(a) The rule for finding the frequency is to multiply the width of the group by the frequency density.
So, for the group 1–8:
   The boundaries are 0.5 and 8.5
   The width = 8.5 – 0.5 = 8
   and the frequency density = 8.5
   So the frequency = 8 × 8.5 = 68

The boundaries of the next group are 8.5 and 14.5
   So the next group is 9–14
   The width = 14.5 – 8.5 = 6
   and the frequency density = 7.5
   So the frequency = 6 × 7.5 = 45

The boundaries of the next group are 14.5 and 24.5
   So the next group is 15–24
   The width = 24.5 – 14.5 = 10
   and the frequency density = 3.5
   So the frequency = 10 × 3.5 = 35

The boundaries of the last group are 24.5 and 29.5
   So the next group is 25–29
   The width = 29.5 – 24.5 = 5
   and the frequency density = 7
   So the frequency = 5 × 7 = 35

So the table is:

Length cm	Frequency
1 – 8	68
9 – 14	45
15 – 24	35
25 – 29	35

(b) The modal group is the group with the highest frequency. The highest frequency is 68. So the limits of the modal group are 1 and 8.

(c) The median group is the group which contains the middle number when the numbers are written in order. It can be found by adding a cumulative frequency column as shown below:

Length cm	Frequency	Cumulative Frequency
1 – 8	68	68
9 – 14	45	113
15 – 24	35	148
25 – 29	35	183

The total of the frequencies is 183.
$$183 \div 2 = 91.5$$
So the median is the 92nd number which, using the table above, will be in the group 9–14.
So the boundaries of the median group are 8.5 and 14.5.

2. The histogram below shows the mass of different objects in kg.

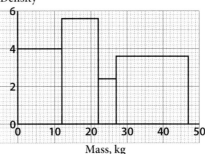

Frequency Density

Mass, kg

(a) Complete the frequency table.

Mass, kg	Frequency
$0 < M \le 12$	

(b) What are the limits of the modal group?
(c) What are the limits of the median group?

(a) To find the frequencies:

For the group $0 < M \leq 12$
   The limits are 0 and 12
   The width = 12 − 0 = 12
   and the frequency density = 4
   So the frequency = 12 × 4 = 48

For the group $12 < M \leq 22$
   The limits are 12 and 22
   The width = 22 − 12 = 10
   and the frequency density = 5.6
   So the frequency = 10 × 5.6 = 56

For the group $22 < M \leq 27$
   The limits are 22 and 27
   The width = 27 − 22 = 5
   and the frequency density = 2.4
   So the frequency = 5 × 2.4 = 12

For the group $27 < M \leq 47$
   The limits are 27 and 47
   The width = 47 − 27 = 20
   and the frequency density = 3.6
   So the frequency = 20 × 3.6 = 72

So the table is:

Mass kg	Frequency
$0 < M \leq 12$	48
$12 < M \leq 22$	56
$22 < M \leq 27$	12
$27 < M \leq 47$	72

(b) The modal group is the group with the highest frequency. The highest frequency is 72 and so the limits of the modal group are 27 and 47.

(c) The median group is the group which contains the middle number when the numbers are written in order. It can be found by adding a cumulative frequency column as shown below

Mass kg	Frequency	Cumulative Frequency
$0 < M \leq 12$	48	48
$12 < M \leq 22$	56	104
$22 < M \leq 27$	12	116
$27 < M \leq 47$	72	188

The total of the frequencies is 188.
$$188 \div 2 = 94$$
So the median is the 94th number which, from the table above, will be in the group $12 < M \leq 22$.
So the limits of the median group are 12 and 22.

Exercise 23D:

1. The histogram below shows the heights of different objects in cm.

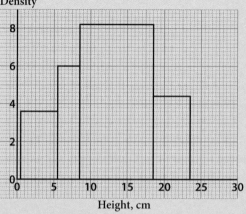

(a) Complete the frequency table:

Length cm	Frequency
1 – 5	

(b) What are the limits of the modal group?
(c) What are the boundaries of the median group?

2. The histogram below shows the volumes, V litres, of different objects.

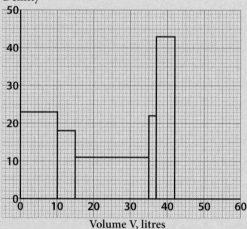

(a) Complete the frequency table:

Volume V, litres	Frequency
$0 < V \leq 10$	

Exercise 23D...

(b) What are the limits of the modal group?

(c) What are the limits of the median group?

3. The histogram below shows the areas, A m², of different objects.

Frequency Density

Area A, cm²

(a) Complete the frequency table:

Area A, m²	Frequency
0 < A≤ 8	

(b) What are the limits of the modal group?

(c) What are the limits of the median group?

4. The histogram below shows the widths of different objects in cm.

Frequency Density

Width, cm

(a) Complete the frequency table:

Width, cm	Frequency
1 – 10	

(b) What are the limits of the modal group?

(c) What are the boundaries of the median group?

5. The histogram below shows the masses of different objects in kg.

Frequency Density

Mass, kg

(a) Complete the frequency table:

Mass, kg	Frequency
1 – 4	

(b) What are the limits of the modal group?

(c) What are the boundaries of the median group?

Exercise 23D...

6. The histogram below shows the lengths, l cm, of different objects.

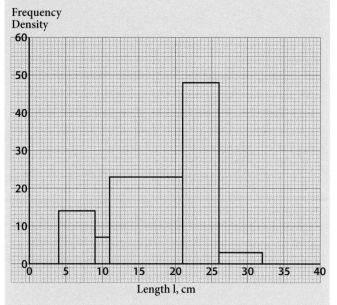

Length l, cm

(a) Complete the frequency table:

Length l, cm	Frequency
4 < l ≤ 9	

(b) What are the limits of the modal group?
(c) What are the limits of the median group?

23.6 Histograms – More Complex Questions

Sometimes you are asked more complex questions, such as ones where the histogram or the data is incomplete. For example:

The frequencies of some of the capacities of containers in litres are shown in the table below:

Capacity C, litres	Frequency
0 < C ≤ 8	32
8 < C ≤ 13	32
13 < C ≤ 33	46
33 < C ≤ 39	
39 < C ≤ 42	

Part of the histogram is drawn below.

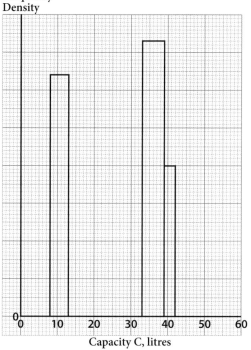

Capacity C, litres

(a) Complete the table.
(b) Complete the histogram.

(a) You must first work out the scale on the vertical axis (frequency density).
For the group 8 < C ≤ 13:
    The limits are 8 and 13
    The width is 13 – 8 = 5
    The frequency is 32 (from the table)
    So the frequency density = 32 ÷ 5 = 6.4
    You can then mark 6.4 on the vertical axis (the frequency density). The vertical axis can then be scaled.
    In this case, 1 cm represents a frequency density of 1.
For the group 33 < C ≤ 39:
    The limits are 33 and 39
    The width is 39 – 33 = 6
    The frequency density is 7.5 (from the histogram)
    So the frequency = 6 × 7.5 = 45
For the group 39 < C ≤ 42:
    The limits are 39 and 42
    The width is 42 – 39 = 3
    The frequency density is 4 (from the histogram)
    So the frequency = 3 × 4 = 12

So the completed table is:

Capacity C, litres	Frequency
0 < C ≤ 8	32
8 < C ≤ 13	32
13 < C ≤ 33	46
33 < C ≤ 39	45
39 < C ≤ 42	12

(b) To complete the histogram, we need to know the frequency density for the two groups not yet drawn.

For the group $0 < C \leq 8$:

The limits are 0 and 8

The width is $8 - 0 = 8$

The frequency density is $32 \div 8 = 4$

For the group $13 < C \leq 33$:

The limits are 13 and 33

The width is $33 - 13 = 20$

The frequency density is $46 \div 20 = 2.3$

The completed histogram is drawn below:

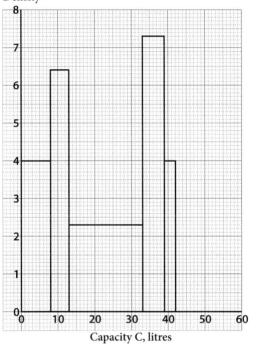

Frequency Density

Capacity C, litres

Exercise 23E:

1. The frequencies of some of the masses of objects in kg are shown in the table below.

Mass, kg	Frequency
5 – 11	119
12 – 16	120
17 – 20	
21 – 30	160
31 – 36	

Exercise 23E...

Part of the histogram is drawn below.

Frequency Density

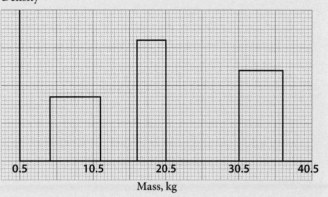

Mass, kg

(a) Complete the table.

(b) Copy and complete the histogram.

2. The frequencies of some of the circumferences of circles in cm are shown in the table below.

Circumference, cm	Frequency
$10 < C \leq 14$	14
$14 < C \leq 19$	42
$19 < C \leq 29$	
$29 < C \leq 31$	9
$31 < C \leq 36$	

Part of the histogram is drawn below.

Frequency Density

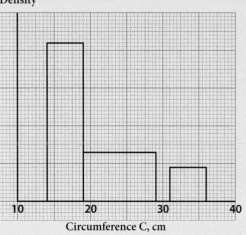

Circumference C, cm

(a) Complete the table.

(b) Copy and complete the histogram.

3. The frequencies of some of the lengths of objects in m are shown in the table below.

Length, m	Frequency
3 – 9	
10 – 14	18
15 – 24	42
25 – 28	30
29 – 33	

Part of the histogram is drawn below.

Frequency
Density

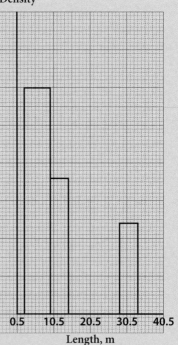

Length, m

(a) Complete the table.
(b) Copy and complete the histogram.

4. The frequencies of some of the perimeters of shapes in cm are shown in the table below.

Perimeter, cm	Frequency
12–16	29
17–20	
21–22	13
23–32	23
33–37	

Part of the histogram is drawn below.

Frequency
Density

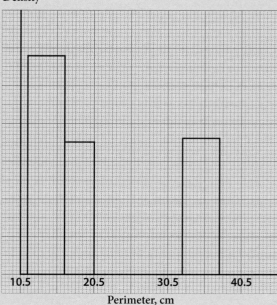

Perimeter, cm

(a) Complete the table.
(b) Copy and complete the histogram.

5. The frequencies of some of the volumes of objects in cm³ are shown in the table below.

Volume V, cm³	Frequency
$5 < V \leq 10$	42
$10 < V \leq 20$	37
$20 < V \leq 24$	
$24 < V \leq 29$	21
$29 < V \leq 33$	

Part of the histogram is drawn overleaf.

Exercise 23E...

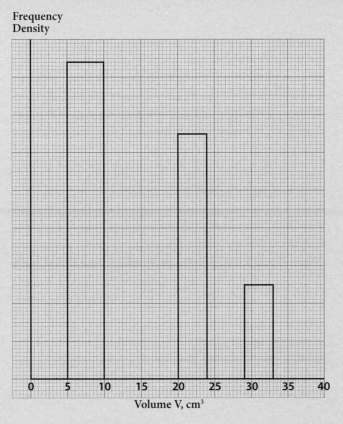

(a) Complete the table.
(b) Copy and complete the histogram.

CHAPTER 24: MEASURES OF CENTRAL TENDENCY AND DISPERSION

24.1 Mean And Standard Deviation Of A List Of Numbers

The **mean** is one of three different statistical averages, the others being the **median** and the **mode**.
The **standard deviation** is a measure of how spread the data is. The standard deviation uses all the data.

The formulae for calculating the mean and standard deviation of a list of numbers are:

Mean
$$\overline{x} = \frac{\Sigma x}{n}$$

Standard Deviation $\quad s = \sqrt{\dfrac{\Sigma x^2}{n} - (\overline{x})^2}$

where
n = the number of figures in the list
Σx = the sum of all the x's
Σx^2 = the sum of all the x^2's

For example:
Find the mean and standard deviation of
6.4, 3.8, 4.7, 5.3 and 4.8

The rules for answering this question are as follows.
For the Mean:

- Add up all the numbers.
- Divide by how many numbers there are.

For the Standard Deviation:

- Square all the numbers.
- Add up the squares.
- Divide the total of the squares by how many numbers there are.
- Subtract the square of the mean.
- Take the square root of your answer.

So we have:

x	x^2
6.4	40.96
3.8	14.44
4.7	22.09
5.3	28.09
4.8	23.04
Total = 25	Total = 128.62

Mean $\quad\quad \overline{x} = \dfrac{\Sigma x}{n}$

$\quad\quad\quad\quad = \dfrac{25}{5} = 5$

Standard Deviation $s = \sqrt{\dfrac{\Sigma x^2}{n} - (\overline{x})^2}$

$\quad\quad\quad\quad = \sqrt{\dfrac{128.62}{5} - 5^2}$

$\quad\quad\quad\quad = \sqrt{0.724}$

$\quad\quad\quad\quad = 0.851$

Exercise 24A:

Find the mean and standard deviation of the following

1. 14, 11, 16, 13

2. 9.4, 6.8, 5.2, 7.3, 4.9

3. 127, 114, 138, 121, 134

4. £5.24, £3.66, £9.24, £4.54

5. 15, 11, 14, 12, 10, 9, 8, 12

6. 7, 11, 13, 8, 6, 7, 9, 5

7. 18, 22, 23, 25, 26, 27, 28, 29

8. 14, 15, 17, 19, 21, 22, 22, 23, 24, 26

24.2 Mean And Standard Deviation Of A Frequency Distribution

The formulae for calculating the mean and standard deviation of a frequency distribution are:

Mean $\quad\quad \overline{x} = \dfrac{\Sigma fx}{n}$

Standard Deviation $\quad s = \sqrt{\dfrac{\Sigma fx^2}{n} - (\overline{x})^2}$

where $\quad\quad$ n = the number of figures in the list, or Σf

$\quad\quad\quad\quad \Sigma fx$ = the sum of all the fx's

$\quad\quad\quad\quad \Sigma fx^2$ = the sum of all the fx^2's

For example:

Find the mean and standard deviation of the frequency distribution below:

x	f
1	2
2	7
3	4
4	5

The rules for answering this question are as follows. For the Mean:
- Multiply f by x to get fx for each number.
- Add up all the fx's.
- Divide Σfx by the sum of the frequencies.

For the Standard Deviation:
- Multiply fx by x to get fx^2 for each number.
- Add up all the fx^2.
- Divide Σfx^2 by the sum of the frequencies.
- Subtract the square of the mean.
- Take the square root of your answer.

So we have:

x	f	fx	fx^2
1	2	2	2
2	7	14	28
3	4	12	36
4	5	20	80
	Total = 18	Total = 48	Total = 146

Mean $\quad\quad \overline{x} = \dfrac{\Sigma fx}{n}$

$\quad\quad\quad\quad = \dfrac{48}{18} = \dfrac{8}{3}$ or $2\dfrac{2}{3}$

Standard Deviation $s = \sqrt{\dfrac{\Sigma fx^2}{n} - (\overline{x})^2}$

$\quad\quad\quad\quad = \sqrt{\dfrac{146}{18} - (2\dfrac{2}{3})^2}$

$\quad\quad\quad\quad = \sqrt{1}$

$\quad\quad\quad\quad = 1$

Exercise 24B: *Find the mean and standard deviation of the following*

1.

x	f
1	6
2	9
3	4
4	1

2.

x	f
2	7
4	11
6	10
8	2

3.

x	f
0	6
3	9
6	8
9	2

4.

x	f
1	16
2	22
3	7
4	5

5. The number of children in different families is shown below. Find the mean and standard deviation.

Number of children	Frequency
0	7
1	4
2	9
3	8
4	5
5	3
6	4

6. The scores in mental maths tests are shown below. Find the mean and standard deviation.

Score	Frequency
2	3
3	3
4	2
5	5
6	4
7	1
8	2
9	4
10	1

7. The number of children in each class in school is shown below. Find the mean and standard deviation.

Number of children in each class	Frequency
23	4
24	3
25	5
26	8
27	2
28	4
29	5
30	1

8. The ages of children in Sunday school are shown below. Find the mean and standard deviation.

Age (years)	Frequency
5	12
6	17
7	13
8	11
9	12
10	11
11	12
12	12

9. The weights of objects to the nearest kg are shown below. Find the mean and standard deviation.

Weight (kg)	Frequency
4	3
5	7
6	9
7	12
8	14
9	5

10. The number of text messages sent by teenagers in one day are shown below. Find the mean and standard deviation.

Number of text messages	Frequency
12	8
13	11
14	6
15	2
16	4
17	9

24.3 Mean And Standard Deviation Of A Grouped Frequency Distribution

If you have a grouped frequency distribution, you need do an additional step:

- Take **x** to be the mid-value of each group.
- Repeat process used with the frequency distribution.

You can work out the mid-value of a group by adding the limits and then dividing by 2. Thus the mid-value of the group 15–29 is 22 since 15 + 29 = 44 and 44 ÷ 2 = 22. Note that a calculation of the mean or standard deviation from a grouped distribution list is only an **estimate** because the original data is not known.

For example:

Find the estimate of the mean and standard deviation of the grouped frequency distribution below:

Group	Frequency
1 < M ≤ 4	7
4 < M ≤ 10	4
10 < M ≤ 11	9
11 < M ≤ 18	3

Taking x to be the mid-values gives the following:

Group	f	x	fx	fx²
1 < M ≤ 4	7	2.5	17.5	43.75
4 < M ≤ 10	4	7	28	196
10 < M ≤ 11	9	10.5	94.5	992.25
11 < M ≤ 18	3	14.5	43.5	630.75
	Total = 23		Total = 183.5	Total = 1862.75

Mean $\quad \overline{x} = \dfrac{\Sigma fx}{n}$

$\qquad = \dfrac{183.5}{23} = 7.98$ to 3 significant figures

Standard Deviation $s = \sqrt{\dfrac{\Sigma fx^2}{n} - (\overline{x})^2}$

$\qquad = \sqrt{\dfrac{1862.75}{23} - 7.98^2}$

$\qquad = \sqrt{17.3} = 4.16$

Exercise 24C: *Find the estimate of the mean and standard deviation of the following*

1.

Group	Frequency
1 < l ≤ 6	7
6 < l ≤ 11	9
11 < l ≤ 16	4
16 < l ≤ 21	5

Exercise 24C...

2.

Group	Frequency
0 – 8	14
9 – 16	21
17 – 20	11
21 – 25	4

3.

Group	Frequency
2 < w ≤ 4	3
4 < w ≤ 6	8
6 < w ≤ 8	4
8 < w ≤ 10	5

4. The marks of pupils in a test are shown below. Find the mean and standard deviation.

Marks	Frequency
1 – 10	4
11 – 20	8
21 – 30	9
31 – 40	14
41 – 50	15

5. The ages of people on an excursion are shown below. Find the mean and standard deviation.

Age (years)	Frequency
1 – 5	8
6 – 10	14
11 – 15	9
16 – 20	10
21 – 25	9

6. The lengths of objects in cm to the nearest cm are shown below. Find the mean and standard deviation.

Length (cm)	Frequency
1 – 5	13
6 – 8	11
9 – 16	7
17 – 20	9

7. The heights of of objects in m to the nearest m are shown below. Find the mean and standard deviation.

Height (m)	Frequency
1 – 9	8
10 – 14	15
15 – 20	6
21 – 30	4

8. The lengths of objects in cm to the nearest cm are shown below. Find the mean and standard deviation.

Length (l cm)	Frequency
0 < l ≤ 10	7
10 < l ≤ 12	9
12 < l ≤ 20	14
20 < l ≤ 24	10

9. The times taken to complete a race are shown below. Find the mean and standard deviation.

Time (t mins)	Frequency
2 < t ≤ 5	16
5 < t ≤ 9	22
9 < t ≤ 14	14
14 < t ≤ 16	8

10. The heights of of objects in m to the nearest m are shown below. Find the mean and standard deviation.

Height (m)	Frequency
1 – 5	7
6 – 10	4
11 – 15	12
16 – 20	18
21 – 25	9

11. The widths of shapes are shown below. Find the mean and standard deviation.

Width (w cm)	Frequency
0 < w ≤ 10	3
10 < w ≤ 15	4
15 < w ≤ 20	8
20 < w ≤ 24	12
24 < w ≤ 30	10
30 < w ≤ 40	3

12. The volumes of shapes are shown below. Find the mean and standard deviation.

Volume (V litres)	Frequency
1 < V ≤ 8	3
8 < V ≤ 12	16
12 < V ≤ 18	22
18 < V ≤ 20	34
20 < V ≤ 27	24
27 < V ≤ 40	1

13. The masses of objects in kg to the nearest kg are shown below. Find the mean and standard deviation.

Mass (kg)	Frequency
1 – 7	6
8 – 14	11
15 – 21	14
22 – 28	9
29 – 35	10

14. The areas of fields in hectares are shown below. Find the mean and standard deviation.

Area (A hectares)	Frequency
1 < A ≤ 9	3
9 < A ≤ 17	5
17 < A ≤ 25	4
25 < A ≤ 33	6
33 < A ≤ 41	2

24.4 Harder Problems

Sometimes you are given incomplete data with which to work. For example:

The frequency table below shows the marks in a test.

Marks	Frequency
1 – 5	8
6 – 7	
8 – 13	12
14 – 16	12
17 – 22	4

The mean mark is 9.98. Find:
(i) the frequency for the 6 – 7 group
(ii) the standard deviation

(i) To solve this problem you must:
• Call the missing frequency n.
• Work out the mid-values of each group (x).
• Multiply the mid-values by the frequency to get fx.
• Add up all the fx's.
• Divide Σfx by the sum of the frequencies to get the mean.
• Form an equation and solve it to find the missing frequency.

So we have:

Mark	Frequency f	Mid-value x	fx
1 – 5	8	3	24
6 – 7	**n**	6.5	6.5n
8 – 13	12	10.5	126
14 – 16	12	15	180
17 – 22	4	19.5	78
	Total = 36 + n		Total = 408 + 6.5n

The mean $\bar{x} = \dfrac{\Sigma fx}{n}$

$$= \frac{408 + 6.5n}{36 + n} = 9.98$$

Cross multiplying gives:

$$9.98(36 + n) = 408 + 6.5n$$
$$359.28 + 9.98n = 408 + 6.5n$$
$$9.98n - 6.5n = 408 - 359.28$$
$$3.48n = 48.72$$
$$n = \frac{48.72}{3.48} = 14$$

So the missing frequency n = 14

(ii) You can now replace n with 14 and find the standard deviation as before. So we have:

Mark	Frequency f	Mid-value x	fx	fx^2
1 – 5	8	3	24	72
6 – 7	**14**	6.5	91	591.5
8 – 13	12	10.5	126	1323
14 – 16	12	15	180	2700
17 – 22	4	19.5	78	1521
	Total = 50			Total = 6207.5

Standard Deviation $s = \sqrt{\dfrac{\Sigma fx^2}{n} - (\bar{x})^2}$

$$= \sqrt{\frac{6207.5}{50} - 9.98^2}$$
$$= \sqrt{24.5496}$$
$$= 4.95$$

Exercise 24D:

1. The frequency table below shows the lengths of objects in cm.

Length, cm	Frequency
2 – 5	3
6 – 15	7
16 – 18	
19 – 23	4
24 – 36	15

The mean length is 20.125 cm. Find:
(i) the missing frequency
(ii) the standard deviation.

2. The frequency table below shows the amount spent in £'s by shoppers

Amount spent, £	Frequency
35 – 44	4
45 – 50	17
51 – 60	36
61 – 80	
81 – 84	9

The mean amount spent is £58.6625. Find:
(i) the missing frequency
(ii) the standard deviation.

3. The frequency table below shows the amount of pocket money in £'s for pupils at a school.

Pocket money, £	Frequency
2 – 2.60	3
2.61 – 3.39	7
3.40 – 5	14
5.01 – 5.99	
6 – 10	14

The mean amount of pocket money is £5.294. Find:
(i) the missing frequency
(ii) the standard deviation.

4. The frequency table below shows the widths of objects in cm.

Width, cm	Frequency
2 – 6	
7 – 11	14
12 – 14	11
15 – 18	9
19 – 26	12

The mean width is 14.07 cm. Find:
(i) the missing frequency
(ii) the standard deviation

5. The frequency table below shows the lengths of journeys in km.

Length, km	Frequency
10 – 30	17
31 – 35	24
36 – 40	16
41 – 50	
51 – 70	18

The mean length is 39.665 km. Find:
(i) the missing frequency
(ii) the standard deviation.

6. The frequency table below shows the masses of objects in kg.

Mass, kg	Frequency
1 – 5	
6 – 8	14
9 – 13	9
14 – 16	22

The mean mass is 10.84 kg. Find:
(i) the missing frequency
(ii) the standard deviation.

7. The frequency table below shows the lengths of objects, l, in m.

Length l, m	Frequency
$1 < l \le 6$	8
$6 < l \le 12$	4
$12 < l \le 23$	
$23 < l \le 30$	9

The mean length is 14.9 m. Find:
(i) the missing frequency
(ii) the standard deviation.

8. The perimeters of shapes, P, in cm are shown below.

Perimeter P, cm	Frequency
$5 < P \le 8$	7
$8 < P \le 14$	4
$14 < P \le 26$	3
$26 < P \le 30$	
$30 < P \le 32$	8

The mean perimeter is 19.26 cm. Find:
(i) the missing frequency
(ii) the standard deviation.

9. The frequency table below shows the capacities of objects in ml.

Capacity, ml	Frequency
1 – 80	17
81 – 120	
121 – 200	14
201 – 240	13

The mean capacity is 128.1 ml. Find:
(i) the missing frequency
(ii) the standard deviation.

24.5 Further Problems

Sometimes you know only some of the information. For example:

1. The lengths of 9 objects have mean 8.4 cm and standard deviation 1.6 cm. A tenth object is 9.5 cm long

Find (i) the mean and (ii) standard deviation of all ten objects.

(i) To find the mean, you use these rules:
- Find the total of the 9 objects.
- Add the tenth length.
- Find the mean of the 10 lengths.

Total of 9 objects = $9 \times 8.4 = 75.6$

Total of 10 objects = $75.6 + 9.5 = 85.1$

Mean of 10 objects = $85.1 \div 10 = 8.51$ cm

(ii) To find the standard deviation, you use these rules:
- Find the Σx^2 for the 9 lengths.
- Add on the square of the tenth length.
- Find the standard deviation of the 10 lengths.

So, for the nine objects:

Standard Deviation s = $\sqrt{\dfrac{\Sigma x^2}{n} - (\overline{x})^2}$

$$1.6 = \sqrt{\frac{\Sigma x^2}{9} - 8.4^2}$$

Squaring both sides gives:

$$2.56 = \frac{\Sigma x^2}{9} - 70.56$$

So $\quad \dfrac{\Sigma x^2}{9} = 2.56 + 70.56$

$$\frac{\Sigma x^2}{9} = 73.12$$

Multiplying both sides by 9 gives:

$$\Sigma x^2 = 73.12 \times 9$$
$$= 658.08$$

Then, for the 10 objects:

$$\Sigma x^2 = 658.08 + 9.5^2$$
$$\Sigma x^2 = 748.33$$

So the standard deviation is given by:

$$s = \sqrt{\frac{\Sigma x^2}{n} - (\overline{x})^2}$$
$$= \sqrt{\frac{748.33}{10} - 8.51^2}$$
$$= \sqrt{2.4129}$$
$$= 1.55 \text{ cm}$$

2. The mean and standard deviation of the wages for 25 men in a factory are £176 and £18.40. The mean and standard deviation of the wages for 15 women in the factory are £142 and £20.60. Find (i) the mean and (ii) standard deviation of the wages for all the workers in the factory.

(i) To find the mean, you use these rules:
- Find the total of the men's wages.
- Find the total of the women's wages.
- Find the total of the wages for all the people.
- Find the mean of the wages for all the people.

So, for the men:

$$\text{Total wage} = 176 \times 25 = £4400$$

And for the women:

$$\text{Total wage} = 142 \times 15 = £2130$$

Therefore, for all the workers:

$$\text{Total wage} = £4400 + £2130 = £6530$$

So:

$$\text{Mean wage} = \frac{6530}{40} = £163.25$$

(ii) To find the standard deviation, you use these rules:
- Find the Σx^2 for the men.
- Find the Σx^2 for the women.
- Find the Σx^2 for all the workers.
- Find the standard deviation for all the workers.

So, for the men:

$$\text{Standard Deviation } s = \sqrt{\frac{\Sigma x^2}{n} - (\overline{x})^2}$$
$$18.40 = \sqrt{\frac{\Sigma x^2}{25} - 176^2}$$

Squaring both sides gives:

$$338.56 = \frac{\Sigma x^2}{25} - 30976$$

So

$$\frac{\Sigma x^2}{25} = 338.56 + 30976$$
$$\frac{\Sigma x^2}{25} = 31314.56$$

Multiplying both sides by 25 gives:

$$\Sigma x^2 = 782864$$

Then, for the women:

Standard Deviation $s = \sqrt{\frac{\Sigma x^2}{n} - (\overline{x})^2}$

$$20.60 = \sqrt{\frac{\Sigma x^2}{15} - 142^2}$$

Squaring both sides gives:

$$424.36 = \frac{\Sigma x^2}{15} - 20164$$

So

$$\frac{\Sigma x^2}{15} = 424.36 + 20164$$
$$\frac{\Sigma x^2}{25} = 20588.36$$

Multiplying both sides by 15 gives:

$$\Sigma x^2 = 308825.4$$

Therefore, for all the workers:

$$\Sigma x^2 = 782864 + 308825.4$$
$$= 1091689.4$$

So the standard deviation is given by:

$$s = \sqrt{\frac{\Sigma x^2}{n} - (\overline{x})^2}$$
$$= \sqrt{\frac{1091689.4}{40} - 163.25^2}$$
$$= \sqrt{641.6725}$$
$$= £25.33$$

Exercise 24E:

1. The mean and standard deviation of 8 weights are 46 kg and 4 kg. A ninth weight is 40 kg. Find the mean and standard deviation of all nine weights.

2. The mean and standard deviation of 12 marks in a test are 53 and 8. A pupil who missed the test then scored 56 marks. Find the mean and standard deviation of all 13 marks.

3. There are 18 girls and 10 boys in a class. The mean and standard deviation of the marks of the girls are 46 and 5

Exercise 24E...

The mean and standard deviation of the marks of the boys are 38 and 8. Find the mean and standard deviation of the whole class.

4. A company employs full time and part time staff. There are 24 full time and 10 part time staff. The mean and standard deviation of the wages of the full time staff are £235.60 and £18.40. The mean and standard deviation of the wages of all the employers are £221.47 and £27.87. Find the mean and standard deviation of the part time staff.

5. The lengths of 25 objects were recorded. The mean and standard deviation of these lengths were 5.04 cm and 0.41 cm. A sample of 10 objects was chosen from these 25 objects. The mean and standard deviation of the lengths in the sample were 4.8 cm and 0.3 cm. Find the mean and standard deviation of the lengths of the 15 objects not in the sample.

6. The mean and standard deviation of the marks of 15 boys in a class test were 64 and 7. The mean and standard deviation of the marks of 10 girls in this class test were 68 and 5. Find the mean and standard deviation of the whole class.

7. The mean and standard deviation of the marks of 30 pupils in an exam were 54 and 6. It was then discovered that one of the pupils had cheated and his mark of 47 was then excluded from the data. Find the mean and standard deviation of the 29 pupils.

8. The mean and standard deviation of 20 diameters of circles are 47 cm and 8 cm. The mean and standard deviation of another set of 30 diameters of circles are 52 cm and 6 cm. Find the mean and standard deviation of all 50 diameters.

24.6 Calculating The Median Of A Grouped Distribution

The median of a grouped distribution is calculated using the following formula:

$$\text{Median} = L_1 + \frac{\left\{\frac{N}{2} - (\Sigma f)_1\right\}c}{f_{med}} \quad \text{where:}$$

$L_1 = $ the lower class boundary
$N = $ the total frequency
$(\Sigma f)_1 = $ the sum of the frequencies up to but not including the median class.
$f_{med} = $ the frequency of the median class
$c = $ the width of the median class

For example:

1. Find the median of the following distribution:

Length l, cm	Frequency
$0 < l \le 4$	14
$4 < l \le 8$	32
$8 < l \le 12$	24
$12 < l \le 16$	6

The rules for solving this problem are:
• Set up a cumulative frequency column to help find the class in which the median lies.
• Use the formula to find the median using the limits.

Length l, cm	Frequency	Cumulative Frequency
$0 < l \le 4$	14	14
$4 < l \le 8$	32	46
$8 < l \le 12$	24	70
$12 < l \le 16$	6	76

$\frac{76}{2} = 38$, so the median is the 38½th number.

The 38½th number is in the class $4 < l \le 8$, so we have:
$L_1 = $ 4 (we have to use the lower class limit as there are no boundaries)
$N = $ 76 (the total of all the frequencies)
$(\Sigma f)_1 = $ 14 (which is the frequency of the class before the median class)
$f_{med} = $ 32 (which is the frequency of the median class)
$c = $ $8 - 4 = 4$
Substituting these figures into the formula gives:

$$\text{Median} = L_1 + \frac{\left\{\frac{N}{2} - (\Sigma f)_1\right\}c}{f_{med}}$$

$$= 4 + \frac{(38 - 14) \times 4}{32}$$

$$= 4 + \frac{96}{32} = 4 + 3 = 7$$

2. Find the median of the following distribution.

Width cm	Frequency
1 – 6	6
7 – 9	15
10 – 24	28
25 – 32	4

The rules for solving this problem are:
• Set up a cumulative frequency column to help find the class in which the median lies.
• Use the formula to find the median using the boundaries.

Width cm	Frequency	Cumulative Frequency
1 – 6	6	6
7 – 9	15	21
10 – 24	28	49
25 – 32	4	53

$\frac{53}{2} = 26.5$, so the median is the 27th number.

The 27th number is in the class 10 – 24, so we have:

$L_1 =$ 9.5 (the lower class boundary of the group)

$N =$ 53 (the total of all the frequencies)

$(\Sigma f)_1 = 6 + 15 = 21$ (the sum of the frequencies up to but not including the median class)

$f_{med} =$ 28 (which is the frequency of the median class)

$c =$ 24.5 – 9.5 = 15

Substituting these figures into the formula gives:

$$\text{Median} = L_1 + \frac{\left\{ \frac{N}{2} - (\Sigma f)_1 \right\} c}{f_{med}}$$

$$= 9.5 + \frac{(26.5 - 21) \times 15}{28}$$

$$= 9.5 + \frac{82.5}{28} = 9.5 + 2.95 = 12.45$$

Exercise 24F: *Find the estimate of the median of the following distributions.*

1.

Length l, cm	Frequency
$0 < l \le 6$	8
$6 < l \le 12$	14
$12 < l \le 18$	10
$18 < l \le 24$	22
$24 < l \le 30$	4

2.

Weight, kg	Frequency
1 – 5	15
6 – 10	14
11 – 15	20
16 – 20	12
21 – 25	13

3.

Height h, cm	Frequency
$0 < h \le 3$	6
$3 < h \le 6$	15
$6 < h \le 9$	18
$9 < h \le 12$	7
$12 < h \le 15$	14
$15 < h \le 18$	3

4.

Marks	Frequency
1 – 5	8
6 – 7	14
8 – 13	12
14 – 16	12
17 – 22	4

5.

Length, cm	Frequency
2 – 5	3
6 – 15	7
16 – 18	11
19 – 23	4
24 – 36	15

6.

Amount spent, £	Frequency
35 – 44	4
45 – 50	17
51 – 60	36
61 – 80	14
81 – 84	9

7.

Pocket money, £	Frequency
2 – 2.60	3
2.61 – 3.39	7
3.40 – 5	14
5.01 – 5.99	12
6 – 10	14

8.

Width, cm	Frequency
2 – 6	7
7 – 11	14
12 – 14	11
15 – 18	19
19 – 26	22

9.

Length, km	Frequency
10 – 30	17
31 – 35	24
36 – 40	16
41 – 50	13
51 – 70	18

Exercise 24F...

10.

Mass, kg	Frequency
1 – 5	8
6 – 8	14
9 – 13	9
14 – 16	22

11.

Perimeter P, cm	Frequency
$5 < P \leq 8$	7
$8 < P \leq 14$	4
$14 < P \leq 26$	3
$26 < P \leq 30$	9
$30 < P \leq 32$	8

12.

Capacity, ml	Frequency
1 – 80	17
81 – 120	22
121 – 200	14
201 – 240	13

24.7 Transforming Sets Of Data

Sometimes you are asked to transform data, ie when every figure is modified in some way. For example:

Consider the set of values: 1, 2, 3, 4, 5.
The mean and standard deviation of these figures can be found as follows:

x	x^2
1	1
2	4
3	9
4	16
5	25
Total = 15	Total = 55

Mean $\qquad \bar{x} = \dfrac{\Sigma x}{n}$

$\qquad\qquad\quad = \dfrac{15}{5} = 3$

Standard Deviation s $= \sqrt{\dfrac{\Sigma x^2}{n} - (\bar{x})^2}$

$\qquad\qquad = \sqrt{\dfrac{55}{5} - 3^2}$

$\qquad\qquad = \sqrt{2}$

$\qquad\qquad = 1.414$

What happens to the mean and standard deviation of these figures when
(i) 6 is added to each figure?
(ii) each figure is trebled?
(iii) each figure is divided by 4 and then 2 is subtracted?

(i) Adding 6 to each figure:

x	x^2
7	49
8	64
9	81
10	100
11	121
Total = 45	Total = 415

Mean $\qquad \bar{x} = \dfrac{45}{5} = 9$

Standard Deviation s $= \sqrt{\dfrac{415}{5} - 9^2}$

$\qquad\qquad = \sqrt{2}$

$\qquad\qquad = 1.414$

Note that:
• The mean is **6 more** than the original mean.
• The standard deviation is the **same** as the original standard deviation.

(ii) Trebling each figure:

x	x^2
3	9
6	36
9	81
12	144
15	225
Total = 45	Total = 495

Mean $\qquad \bar{x} = \dfrac{45}{5} = 9$

Standard Deviation s $= \sqrt{\dfrac{495}{5} - 9^2}$

$\qquad\qquad = \sqrt{18}$

$\qquad\qquad = 3\sqrt{2}$

$\qquad\qquad = 4.24$

Note that:
• The mean is **3 times** the original mean.
• The standard deviation is **3 times** the original standard deviation.

(iii) Dividing each figure by 4 and then subtracting 2

x	x^2
−1.75	3.0625
−1.5	2.25
−1.25	1.5625
−1	1
−0.75	0.5625
Total = −6.25	Total = 8.4375

Mean $\qquad \overline{x} = \dfrac{-6.25}{5} = -1.25$

Standard Deviation s $= \sqrt{\dfrac{8.4375}{5} - (-1.25)^2}$

$\qquad\qquad = \sqrt{0.125}$

$\qquad\qquad = \frac{1}{4}\sqrt{2}$

Note that:
• The mean is the original mean **divided by 4 and then 2 subtracted**.
• The standard deviation is the original standard deviation **divided by 4**.

The above examples show the following rules for transforming data.
For changes to the **mean**, if the mean of a set of data is $\overline{x}$, then:
• Adding y to all the figures changes the mean to $\overline{x} + y$
• Subtracting y from all the figures changes the mean to $\overline{x} - y$
• Multiplying all the figures by y changes the mean to $y\overline{x}$
• Dividing all the figures by y changes the mean to $\dfrac{\overline{x}}{y}$

For changes to the **standard deviation**, if the standard deviation of a set of data is s, then:
• Adding y to all the figures leaves the standard deviation unchanged
• Subtracting y from all the figures leaves the standard deviation unchanged
• Multiplying all the figures by y changes the standard deviation to ys
• Dividing all the figures by y changes the standard deviation to $\dfrac{s}{y}$

For example:
The scores in a test have mean 54 and standard deviation 5.6. Find the new mean and new standard deviation when all the figures are:
(i) increased by 7
(ii) halved
(iii) tripled and then decreased by 7

(i)
New Mean = 54 + 7 = 61
New Standard Deviation = 5.6
(ii)
New Mean = $\dfrac{54}{2}$ = 27
New Standard Deviation = $\dfrac{5.6}{2}$ = 2.8
(iii)
New Mean = 54 × 3 − 7 = 155
New Standard Deviation = 5.6 × 3 = 16.8

Exercise 24G:

1. The scores in a test have mean 46 and standard deviation 4.7. Find the new mean and new standard deviation when all the figures are:
   (i) increased by 6
   (ii) doubled
   (iii) halved and then 3 subtracted

2. The heights of objects have mean 15.2 cm and standard deviation 2.8 cm. Find the new mean and new standard deviation when all the figures are:
   (i) multiplied by 4
   (ii) reduced by 5
   (iii) halved and then 7 added

3. The ages of people in a club have mean 54 and standard deviation 2.9. Find the new mean and new standard deviation when all the years are:
   (i) divided by 4
   (ii) increased by 10
   (iii) multiplied by 5 and then reduced by 2

4. The times that substances take to dissolve were recorded with mean 1.4 minutes and standard deviation 0.24 minutes. Find the new mean and new standard deviation in minutes when all the times are:
   (i) reduced by 30 seconds
   (ii) tripled
   (iii) halved and then increased by 15 seconds

5. The lengths of a set of objects have mean 5.2 m and standard deviation 1.2 m. Find the new mean and new standard deviation in metres when all the lengths are:
   (i) increased by 60 cm
   (ii) halved
   (iii) tripled and then reduced by 20 cm

Exercise 24G...

6. The radii of a set of circles have mean 8.4 cm and standard deviation 2.31 cm. Find the new mean and new standard deviation when all the radii are:

   (i) reduced by 2.5 cm
   (ii) multiplied by 4
   (iii) divided by 3 and then increased by 1.4 cm

7. The IQs of people in a club were measured. The mean IQ was 114 and the standard deviation of the IQs was 18. Find the new mean and new standard deviation when all the IQs were:

   (i) increased by 23
   (ii) divided by 3
   (iii) doubled and then reduced by 15

8. The distances pupils walk to school have mean 2.8 km and standard deviation 0.3 km. Find the new mean and new standard deviation in km when all the distances are:

   (i) reduced by 600 m
   (ii) doubled
   (iii) divided by 4 and then increased by 200 m

9. The goals scored in a football tournament have mean 2.43 and standard deviation 0.27. Find the new mean and new standard deviation when all the goals are:

   (i) increased by 5
   (ii) divided by 9
   (iii) tripled and then reduced by 4

10. The capacities of a set of liquids have mean 3.4 l and standard deviation 0.8 l. Find the new mean and new standard deviation in litres when all the capacities are:

   (i) reduced by 800 ml
   (ii) doubled
   (iii) divided by 5 and then increased by 150 ml

CHAPTER 25: PROBABILITY

25.1 The Rules of Probability

Probability is the study of how likely it is that one or more events will happen. A probability is expressed as a value between 0 (definitely won't happen) and 1 (definitely will happen). Here are three rules of probability that you need to know:

The AND Rule For Independent Events

Two events A and B are **independent** if the outcome of A does not affect the outcome of B. If P(A) is the probability that event A happens and P(B) is the probability that event B happens and A and B are independent then:

$$P(A \text{ and } B) = P(A) \times P(B)$$

The OR Rule For Mutually Exclusive Events

Two events A and B are **mutually exclusive** if A and B cannot occur at the same time. If P(A) is the probability that event A happens and P(B) is the probability that event B happens and A and B are mutually exclusive then:

$$P(A \text{ or } B) = P(A) + P(B)$$

The NOT Rule

If P(A) is the probability that event A happens then:
$$P(\text{not } A) = 1 - P(A)$$

25.2 Tree Diagrams

A tree diagram can be used to work out probabilities. For example:

1. A box contains 8 red and 6 blue marbles. 3 marbles are selected at random one after the other without replacement.

(a) Draw a tree diagram to show all the possible outcomes.

(b) Hence find the probability that:
   (i) the first 3 marbles selected are red, blue and blue in that order.
   (ii) 2 red and 1 blue are selected.
   (iii) at least 1 blue is selected.

(a) When drawing a tree diagram, remember that the probabilities on each branch of the tree diagram must always add up to 1. In this example, it is easier to express the probabilities as fractions. As we are not replacing the marbles, so the first set of branches is expressed as a probability out of 14 marbles, the second out of 13 marbles and the third out of 12 marbles. Hence we have:

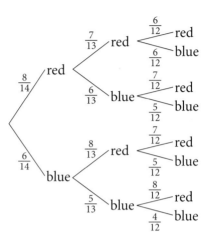

(b) (i) Using the AND rule:

$$P(r \text{ and } b \text{ and } b) = \frac{8}{14} \times \frac{6}{13} \times \frac{5}{12}$$
$$= \frac{10}{91}$$

(ii) You should write out all the possible combinations first. Thus we have

R and R and B or
R and B and R or
B and R and R

You should then work out the probability for each of these combinations by multiplying (AND rule):

$$P(R \text{ and } R \text{ and } B) = \frac{8}{14} \times \frac{7}{13} \times \frac{6}{12}$$
$$= \frac{2}{13}$$
$$P(R \text{ and } B \text{ and } R) = \frac{8}{14} \times \frac{6}{13} \times \frac{7}{12}$$
$$= \frac{2}{13}$$
$$P(B \text{ and } R \text{ and } R) = \frac{6}{14} \times \frac{8}{13} \times \frac{7}{12}$$
$$= \frac{2}{13}$$

Because you want the probability that **any** of these combinations will happen, you should add these probabilities (OR rule) to get:

$$P(2 \text{ red and } 1 \text{ blue}) = \frac{2}{13} + \frac{2}{13} + \frac{2}{13}$$
$$= \frac{6}{13}$$

(iii) It is quicker to find the probability that no blues are selected and then subtract this probability from 1 (ie, using the NOT rule) to find the probability that at least 1 blue is selected.

$$P(\text{no blues}) = P(\text{red and red and red})$$
$$= \frac{8}{14} \times \frac{7}{13} \times \frac{6}{12}$$
$$= \frac{2}{13}$$

So $P(\text{at least 1 blue is selected}) = 1 - \frac{2}{13} = \frac{11}{13}$

2. The probability it is sunny is 0.74. If it is sunny then the probability that Yasmin goes for a walk is 0.85. Otherwise the probability that Yasmin goes for a walk is 0.28.

(a) Draw a tree diagram to show all the possible outcomes.
(b) Hence find the probability:
   (i) it is sunny and Yasmin goes for a walk.
   (ii) Yasmin goes for a walk.

(a) As before, we draw the tree diagram showing the probabilities given in the question:

```
          0.85 walk
    0.74  sunny
          0.15 not walk

          0.28 walk
    0.26  not sunny
          0.72 not walk
```

(b) (i) P(it is sunny and Yasmin goes for a walk)
$$= 0.74 \times 0.85 = 0.629$$

(ii) There are two possible scenario that result in Yasmin going for a walk:

P(it is sunny and Yasmin goes for a walk)
$$= 0.74 \times 0.85 = 0.629 \text{ or}$$
P(it is not sunny and Yasmin goes for a walk)
$$= 0.26 \times 0.28 = 0.0728$$

So: P(Yasmin goes for a walk)
$$= 0.629 + 0.0728 = 0.7018$$

Exercise 25A:

1. The probability traffic lights are red is 0.5.
   The probability traffic lights are amber is 0.2.
   The probability traffic lights are green is 0.3.
   A car approaches two sets of traffic lights.
   (a) Draw a tree diagram to show all the possible outcomes.
   (b) Hence find the probability that:
   (i) one set of traffic lights is red and the other set of traffic lights is green.
   (ii) both traffic lights are different.
   (iii) at least one set of traffic lights is green.

Exercise 25A...

2. A box contains 12 green and 18 yellow discs. Two discs are selected at random, one after the other, and replaced each time.
(a) Draw a tree diagram to show all the possible outcomes.
(b) Hence find the probability that:
(i) the first disc is green and the second is yellow.
(ii) both are green.
(iii) one is green and the other is yellow.

3. The probability Maureen eats chips is 0.76. If Maureen eats chips then the probability she eats a burger is 0.6. Otherwise the probability she eats a burger is 0.48.
(a) Draw a tree diagram to show all the possible outcomes.
(b) Hence find the probability that:
(i) she eats chips and a burger.
(ii) she eats a burger.

4. The probability Gary decides to study French is 0.7. If Gary studies French then the probability he decides to study German is 0.28. Otherwise the probability he decides to study German is 0.62.
(a) Draw a tree diagram to show all the possible outcomes.
(b) Hence find the probability that:
(i) he decides to study French and German.
(ii) he decides to study German.

5. There are 15 boys and 12 girls in a class. Three pupils are selected at random.
(a) Draw a tree diagram to show all the possible outcomes.
(b) Hence find the probability that:
(i) a girl and then a boy and then a girl are selected.
(ii) at least one boy is selected.
(iii) more girls are selected.

6. The probability that Veronica owns a dog is 0.81. If Veronica owns a dog then the probability she owns a cat is 0.62. Otherwise the probability she owns a cat is 0.68.
(a) Draw a tree diagram to show all the possible outcomes.
(b) Hence find the probability that:
(i) she doesn't own a dog but she owns a cat.
(ii) she owns a cat.

7. Box A contains six 20p coins and four 50p coins. Box B contains three 20p coins and seven 50p coins. A coin is chosen at random from box A and placed in box B. A coin is then chosen at random from box B.
(a) Draw a tree diagram to show all the possible outcomes.
(b) Hence find the probability that:
(i) a 20p coin is chosen from box B.
(ii) a coin of the same value is chosen each time.

8. The probability Arlene goes to Belfast by train is 0.44. If Arlene goes to Belfast by train then the probability she visits the Titanic Quarter is 0.35. Otherwise the probability she visits the Titanic quarter is 0.42.
(a) Draw a tree diagram to show all the possible outcomes.
(b) Hence find the probability that:
(i) she goes to Belfast by train and she visits the Titanic Quarter.
(ii) she visits the Titanic quarter.

25.3 Venn Diagrams

A Venn diagram uses circles to show how different sets overlap each other. The rules for drawing a Venn diagram with two sets are as follows:

• Begin by filling in the intersection of the two circles.
• Then fill in the other parts of each circle.
• Fill in outside the two circles for those values that are in neither set.

For example:
1. There are 42 girls in Year 13.
   23 play netball
   17 play hockey
   8 play neither
(a) Draw a Venn diagram to show this information.
(b) Hence use the Venn diagram to find how many girls play both netball and hockey.
(c) A girl is chosen at random. Find the probability she
   (i) plays both netball and hockey
   (ii) plays netball but not hockey.

(a) Let x = the number of girls who play both netball and hockey. Thus we can say:
($23 - x$) girls play netball but not hockey, and
($17 - x$) girls play hockey but not netball
We can therefore draw the Venn diagram:

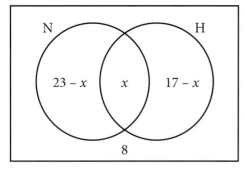

(b) The total number of girls is 42. The four parts of the Venn diagram must add up to this total.

Thus	$42 = 23 - x + x + 17 - x + 8$
So	$42 = 48 - x$
So	$x = 48 - 42$
	$= 6$ girls who play both

(c) (i) There are 42 girls in total. 6 girls play both netball and hockey. So:

P(girl plays both netball and hockey) = $\frac{6}{42} = \frac{1}{7}$

$23 - 6 = 17$ girls play netball but not hockey. So:

(ii) P(girl plays netball but not hockey) = $\frac{17}{42}$

2. There are 35 students in a class:
2 students study Physics, Chemistry and Biology
6 students study Physics and Biology
9 students study Physics and Chemistry
5 students study Biology and Chemistry
18 students study Physics
21 students study Chemistry
10 students study Biology
(a) Draw a Venn diagram to show this information.
(b) A student is chosen at random. Find the probability
  (i) that he/she studies only Physics
  (ii) that he/she studies Chemistry but not Biology.

(a)
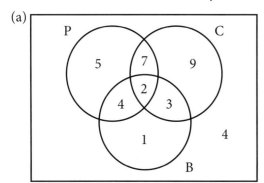

(b) (i) 5 students study only Physics. So:
P(only Physics) = $\frac{5}{35}$ or $\frac{1}{7}$

(ii) From the Venn diagram, the number of students studying Chemistry but not Biology = 7 + 9 = 16. So:

P(Chemistry but not Biology) = $\frac{16}{35}$

Exercise 25B:

1. There are 18 boys in a youth club.
8 play rugby
12 play football
4 play neither
(a) Draw a Venn diagram to show this information.
(b) Hence use the Venn diagram to find how many boys play both rugby and football.
(c) A boy is chosen at random. Find the probability he:
(i) plays both rugby and football
(ii) plays rugby but not football.

2. There are 31 pupils in a form class.
17 study Irish
19 study Spanish
3 study neither
(a) Draw a Venn diagram to show this information.
(b) Hence use the Venn diagram to find how many pupils study both Irish and Spanish.
(c) A pupil is chosen at random. Find the probability he:
(i) studies only Irish
(ii) does not study Spanish.

3. There are 33 teenagers on a trip to a concert.
18 buy a CD
21 buy a DVD
8 buy neither
(a) Draw a Venn diagram to show this information.
(b) Hence use the Venn diagram to find how many teenagers bought both a CD and a DVD.
(c) A teenager is chosen at random. Find the probability she:
(i) buys neither a CD nor a DVD
(ii) buys both a CD and a DVD.

4. There are 55 books on book shelf.
26 are hard back
8 are hard back romantic books
15 are neither hard back nor romantic

Exercise 25B...

(a) Draw a Venn diagram to show this information.
(b) Hence use the Venn diagram to find how many books are:
(i) romantic but not hard back
(ii) romantic.
(c) A book is chosen at random. Find the probability it is:
(i) a hard back
(ii) not a hard back nor romantic.

5. There are 31 people in a youth club.
21 play chess
13 play chess and draughts
4 play neither
(a) Draw a Venn diagram to show this information.
(b) Hence use the Venn diagram to find how many play only draughts.
(c) A person is chosen at random. Find the probability she plays:
(i) both chess and draughts
(ii) only draughts.

6. 61 students go to after school activities.
23 go to Art club
25 go to computer club
22 go to neither
(a) Draw a Venn diagram to show this information.
(b) Hence use the Venn diagram to find how many go to both Art club and computer club.
(c) A student is chosen at random. Find the probability he goes to:
(i) only the Art club
(ii) both Art club and computer club.

7. 36 people go out together for a meal
22 have a starter
22 have a dessert
3 have neither
(a) Draw a Venn diagram to show this information.
(b) Hence use the Venn diagram to find how many have both a starter and a dessert
(c) A person is chosen at random. Find the probability she has:
(i) a starter
(ii) both a starter and a dessert.

8. There are 27 pupils in a class.
13 wear glasses
7 are left–handed
4 are left–handed and wear glasses.
(a) Draw a Venn diagram to show this information.

(b) Hence use the Venn diagram to find how many pupils are not left–handed and do not wear glasses.
(c) A pupil is chosen at random. Find the probability he:
(i) is left–handed and wear glasses
(ii) wears glasses but is not left–handed
(iii) is left–handed but does not wear glasses.

9. 32 people were asked how they drink coffee.
21 said they took milk with their coffee
8 said they took milk and sugar with their coffee
7 said they took neither milk nor sugar with their coffee.
(a) Draw a Venn diagram to show this information.
(b) Hence use the Venn diagram to find how many took sugar with their coffee .
(c) One of these people was chosen at random. Find the probability she took:
(i) milk and sugar with her coffee
(ii) sugar but not milk with her coffee
(iii) neither milk nor sugar with her coffee.

10. 60 teenagers were asked how they helped at home.
24 said they tidied up their rooms
23 said they helped wash the dishes
22 said they neither tidied up their rooms nor helped wash the dishes.
(a) Draw a Venn diagram to show this information.
(b) Hence use the Venn diagram to find how many tidied up their rooms and helped wash the dishes.
(c) One of these people was chosen at random. Find the probability he:
(i) tidied up his room and helped wash the dishes
(ii) helped wash the dishes but did not tidy up his room
(iii) tidied up his room but did not help wash the dishes.

11. There are 55 teenagers in a youth club:
6 teenagers have a pet dog, cat and rabbit
15 teenagers have a pet dog and cat
10 teenagers have a pet dog and rabbit
9 teenagers have a pet cat and rabbit
33 teenagers have a pet dog
30 teenagers have a pet cat
15 teenagers have a pet rabbit
(a) Draw a Venn diagram to show this information.
(b) A teenager is chosen at random. Find the probability that he or she has:
(i) a rabbit but not a cat
(ii) none of these pets.

Exercise 25B...

12. 37 ladies in a club were discussing where they had been on holiday:

2 had been to Spain, France and Italy
6 had been to Spain and France
5 had been to Spain and Italy
3 had been to France and Italy
20 had been to Spain
14 had been to France
11 had been to Italy

(a) Draw a Venn diagram to show this information.

(b) A lady is chosen at random. Find the probability that she has been to

(i) Spain but not Italy

(ii) France and Italy but not Spain.

25.4 Conditional Probability

Conditional probability is when the probability of the second event is conditional on the first event having happened. The rules for conditional probability are:

P(A|B) means the probability that A will happen given that B has happened:

$$P(A|B) = \frac{P(A \text{ and } B)}{P(B)}$$

P(B|A) means the probability that B will happen given that A has happened:

$$P(B|A) = \frac{P(A \text{ and } B)}{P(A)}$$

For example:

1. The probability a girl in a school plays netball is $\frac{5}{9}$

The probability a girl in a school plays hockey is $\frac{28}{45}$

The probability a girl in a school plays neither netball nor football is $\frac{2}{45}$

(a) Draw a Venn diagram to find the probability that a girl plays both netball and football.

(b) Hence find the probability that a girl chosen at random plays:

   (i) netball, given that she plays hockey

   (ii) hockey, given that she plays netball.

(a)

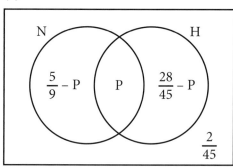

Total of the probabilities inside the Venn diagram

$$= \frac{5}{9} - p + p + \frac{28}{45} - p + \frac{2}{45}$$

$$= 1\frac{2}{9} - p$$

But the total of the probabilities must add up to 1.

So $\qquad 1\frac{2}{9} - p = 1$

Therefore $\qquad p = \frac{2}{9}$

(b) We know:

P(girl plays netball) $= \frac{5}{9}$

P(girl plays hockey) $= \frac{28}{45}$

P(girl plays netball and hockey) $= \frac{2}{9}$

(i) P(netball, given that she plays hockey)

$$= \frac{P(\text{girl plays netball and hockey})}{P(\text{girl plays hockey})}$$

$$= \frac{2}{9} \div \frac{28}{45}$$

$$= \frac{5}{14}$$

(ii) P(hockey, given that she plays netball)

$$= \frac{P(\text{girl plays netball and hockey})}{P(\text{girl plays netball})}$$

$$= \frac{2}{9} \div \frac{5}{9}$$

$$= \frac{2}{5}$$

....

2. The probability a boy chosen at random from Year 11 plays rugby is 0.6. If he plays rugby the probability he plays football is 0.56. Otherwise the probability he plays football is 0.73.

(a) Draw a tree diagram to show all the possible outcomes.

(b) Hence find the probability a boy chosen at random

from Year 11 plays:
   (i) rugby and football
   (ii) football
   (iii) rugby, if he plays football.

(a)

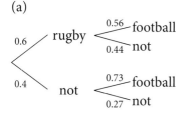

(b) (i) P(boy plays rugby and football)
$$= P(\text{rugby}) \times P(\text{football})$$
$$= 0.6 \times 0.56$$
$$= 0.336$$
(ii) P(boy plays football) = P(boy plays rugby and football) or P(boy does not play rugby and plays football)
$$= 0.336 + (0.4 \times 0.73)$$
$$= 0.628$$
(iii) P(plays rugby if he plays football)
$$= \frac{P(\text{plays rugby and football})}{P(\text{plays football})}$$
$$= 0.336 \div 0.628$$
$$= 0.535$$

3. The probability Alan reads a book is $\frac{2}{7}$

The probability Alan reads a magazine is $\frac{8}{13}$

The probability Alan reads a book and a magazine is $\frac{12}{77}$

Find the probability Alan reads:
(i) a book if he reads a magazine
(ii) a magazine if he reads a book.

(i) P(reads a book if he reads a magazine)
$$= \frac{P(\text{reads a book and a magazine})}{P(\text{reads a magazine})}$$
$$= \frac{12}{77} \div \frac{8}{13}$$
$$= \frac{39}{154}$$

(ii) P(reads a magazine if he reads a book)
$$= \frac{P(\text{reads a book and a magazine})}{P(\text{reads a book})}$$
$$= \frac{12}{77} \div \frac{2}{7}$$
$$= \frac{6}{11}$$

Exercise 25C:

1. 33 pupils go on a sports weekend.
   18 swim
   14 play football
   9 neither swim nor play football
   (a) Draw a Venn diagram to find how many pupils swim and play football.
   (b) Hence find the probability:
   (i) a pupil swims but does not play football
   (ii) a pupil swims given that he plays football
   (iii) a pupil plays football given that he swims.

2. There are 34 pupils in a Year 12 form class.
   13 study History
   22 study Geography
   4 study neither History nor Geography.
   (a) Draw a Venn diagram to find how many pupils study both History and Geography.
   (b) Hence find the probability a pupil
   (i) studies only Geography
   (ii) studies History given that he studies Geography
   (iii) studies Geography given that he studies History.

3. There are 40 cars in a car park.
   23 are foreign made
   7 are black foreign made cars
   8 are neither black nor foreign made cars.
   (a) Draw a Venn diagram to find how many cars are black.
   (b) Hence find the probability a car
   (i) is black
   (ii) is black given that it is foreign made
   (iii) is foreign made given that it is black.

4. There are 39 people in a canteen.
   22 order chips
   20 order ice cream
   4 order neither chips nor ice cream
   (a) Draw a Venn diagram to find how many order chips and ice cream.
   (b) Hence find the probability a person orders
   (i) chips or ice cream
   (ii) chips given that she orders ice cream
   (iii) ice cream given that she orders chips

5. The probability Quinn plays rugby is $\frac{4}{7}$

   The probability Quinn plays snooker is $\frac{2}{5}$

   The probability Quinn plays rugby and snooker is $\frac{1}{3}$

Find the probability Quinn plays
(i) rugby if he plays snooker
(ii) snooker if he plays rugby.

6. The probability it rains is 0.7. If it rains the probability Cadence goes for a swim is 0.84. Otherwise the probability Cadence goes for a swim is 0.35.
(a) Draw a tree diagram to show all the possible outcomes.
(b) Hence find the probability:
(i) it rains and Cadence goes for a swim
(ii) Cadence goes for a swim
(iii) Cadence goes for a swim if it rains
(iv) it rains if Cadence goes for a swim.

7. The probability Lily is late going for a train is 0.64
If Lily is late the probability the train is late is 0.15
Otherwise the probability the train is late is 0.8
(a) Draw a tree diagram to show all the possible outcomes.
(b) Hence find the probability:
(i) Lily is late and the train is late
(ii) the train is late
(iii) Lily is late if the train is late
(iv) the train is late if Lily is late.

8. The probability David revises for his exam is $\frac{7}{9}$

If David revises the probability he passes is $\frac{4}{5}$

Otherwise the probability he passes is $\frac{1}{10}$

(a) Draw a tree diagram to show all the possible outcomes.
(b) Hence find the probability:
(i) David revises and he passes
(ii) David passes
(iii) David passes if he revises.

9. The probability of getting chips for dinner is $\frac{2}{9}$

The probability of getting ice cream for dessert is $\frac{1}{3}$

The probability of getting chips and ice cream is $\frac{1}{30}$

Show that it is more likely to get ice cream given that you get chips rather than to get chips given that you get ice cream.

10. 91 pupils are in Year 12
46 of them went to a rock concert
35 of them went to a country concert
17 of them went to neither.

(a) Draw a Venn diagram to find how many pupils went to both concerts.
(b) Hence find the probability a pupil chosen at random
(i) goes to both a rock concert and a country concert
(ii) goes to the rock concert if she went to the country concert.

11. The probability a car is British built is 0.56
The probability a car is blue is 0.4
The probability a car is a blue British built car is 0.35.
A car is chosen at random. Find the probability:
(i) the car is British built if it is blue
(ii) the car is blue if the car is British built.

12. The probability Sinead passes Maths is 0.6
The probability Sinead passes Further Maths is 0.45
The probability Sinead passes both is 0.35
Find the probability Sinead passes:
(i) Maths if she passes Further Maths
(ii) Further Maths if she passes Maths.

13. The probability Colin catches the flu is $\frac{3}{5}$

The probability Colin develops a chest infection is $\frac{4}{7}$

The probability Colin catches the flu and develops a chest infection is $\frac{4}{11}$

Find the probability Colin:
(i) catches the flu if he develops a chest infection
(ii) develops a chest infection if he catches the flu.

25.5 Harder Questions

Sometimes you are given some probabilities and have to work out others from these. For example:

1. The probability A and B happen is $\frac{8}{27}$

The probability B happens is $\frac{7}{9}$

The probability B happens given that A happens is $\frac{4}{9}$

Find:
(i) the probability A happens given that B happens
(ii) the probability A happens.

(i) P(A happens given that B happens)

$$= \frac{P(A \text{ and } B \text{ happen})}{P(B \text{ happens})}$$

$$= \frac{8}{27} \div \frac{7}{9}$$

$$= \frac{8}{21}$$

(ii) P(B happens given that A happens)

$$= \frac{P(A \text{ and } B \text{ happen})}{P(A \text{ happens})}$$

$$\frac{4}{9} = \frac{8}{27} \div P(A \text{ happens})$$

$$P(A \text{ happens}) = \frac{8}{27} \div \frac{4}{9}$$

$$= \frac{2}{3}$$

2. The probability A happens given that B happens is 0.72
The probability B happens is 0.5
The probability A happens is 0.9
Find:
(i) the probability A and B happen
(ii) the probability B happens given that A happens.

(i) P(A happens given that B happens)

$$= \frac{P(A \text{ and } B \text{ happen})}{P(B \text{ happens})}$$

$$0.72 = \frac{P(A \text{ and } B \text{ happen})}{0.5}$$

$$P(A \text{ and } B \text{ happen}) = 0.72 \times 0.5$$
$$= 0.36$$

(ii) P(B happens given that A happens)

$$= \frac{P(A \text{ and } B \text{ happen})}{P(A \text{ happens})}$$

$$= \frac{0.36}{0.9}$$

$$= 0.4$$

Exercise 25D:

1. The probability A and B happen is 0.24
The probability B happens is 0.3
The probability B happens given that A happens is 0.48
Find:
(i) the probability A happens given that B happens
(ii) the probability A happens.

Exercise 25D...

2. The probability A happens given that B happens is 0.88. The probability B happens is 0.25. The probability A happens is 0.4.
Find:
(i) the probability A and B happen
(ii) the probability B happens given that A happens.

3. The probability A happens given that B happens is 0.32. The probability A and B happen is 0.16. The probability B happens given that A happens is 0.2.
Find:
(i) the probability B happens
(ii) the probability A happens.

4. The probability A and B happen is 0.15.
The probability B happens is 0.3
The probability B happens given that A happens is 0.75. Find:
(i) the probability A happens given that B happens
(ii) the probability A happens.

5. The probability A happens given that B happens is 0.7. The probability A and B happen is 0.42. The probability A happens is 0.8.
Find:
(i) the probability B happens
(ii) the probability B happens given that A happens.

6. The probability A happens given that B happens is $\frac{22}{35}$. The probability B happens is $\frac{4}{11}$.

The probability A happens is $\frac{2}{5}$.
Find:
(i) the probability A and B happen
(ii) the probability B happens given that A happens.

7. The probability A and B happen is 0.049.
The probability B happens is 0.25.
The probability B happens given that A happens is 0.14. Find:
(i) the probability A happens given that B happens
(ii) the probability A happens

8. The probability A happens given that B happens is 0.256. The probability B happens is 0.15. The probability A happens is 0.24. Find:
(i) the probability A and B happen
(ii) the probability B happens given that A happens.

9. The probability A happens given that B happens is 0.896. The probability A and B happen is 0.2688. The probability B happens given that A happens is 0.84. Find:

(i) the probability B happens
(ii) the probability A happens.

10. The probability A and B happen is 0.3648. The probability B happens is 0.48. The probability B happens given that A happens is 0.608. Find:

(i) the probability A happens given that B happens
(ii) the probability A happens.

11. The probability A happens given that B happens is $\frac{5}{12}$. The probability A and B happen is $\frac{5}{32}$.

The probability A happens is $\frac{7}{12}$.
Find:

(i) the probability B happens
(ii) the probability B happens given that A happens.

12. The probability A happens given that B happens is $\frac{7}{12}$.

The probability B happens is $\frac{5}{6}$.

The probability A happens is $\frac{11}{12}$.
Find:

(i) the probability A and B happen
(ii) the probability B happens given that A happens.

13. The probability A happens given that B happens is $\frac{39}{154}$. The probability A and B happen is $\frac{12}{77}$.

The probability B happens given that A happens is $\frac{6}{11}$.
Find:

(i) the probability B happens
(ii) the probability A happens.

CHAPTER 26: BIVARIATE ANALYSIS

26.1 Correlation

Bivariate Analysis is the analysis of two sets of data in order to see if there is a link or connection between them. This link or connection is called the **correlation**. There are two types of correlation:

• **Positive Correlation** is when as one variable increases (or decreases) the other increases (or decreases). This is shown below:

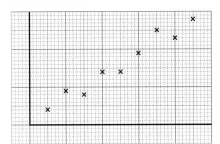

• **Negative Correlation** is when as one variable increases (or decreases) the other decreases (or increases). This is shown on the right:

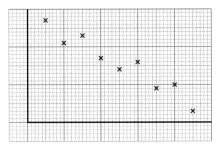

One way to see if there is correlation between two variables is to plot the variables on graph paper as shown above. There is another way to find out if there is correlation. This is by calculating Spearman's Rank Correlation Coefficient as outlined below.

26.2 Spearman's Rank Correlation Coefficient, r

The rules for calculating pearman's Rank Correlation Coefficient are as follows:

• Each piece of data is given a rank depending on its position in the data. You can either rank the smallest piece of data 1, the next smallest piece of data 2 and so on; or you can rank the biggest piece of data 1, the next biggest piece of data 2 and so on.

- Work out the differences, d, between corresponding ranks assigned to each piece of data.
- Calculate Spearman's Rank Correlation Coefficient:

$$r = 1 - \frac{6\Sigma d^2}{n(n^2 - 1)}$$

where: Σd^2 = the sum of all the d^2
            n = the number of pairs of data

You then use Spearman's Rank Correlation Coefficient to determine if there is any correlation , and if so what type of correlation, as follows:

$0 < r \le 1$ indicates positive correlation
  (1 would mean perfect positive correlation)
$-1 \le r < 0$ indicates negative correlation
  (1 would mean perfect negative correlation)

For example:
The table below shows pairs of data.
(i) Calculate Spearman's Rank Correlation Coefficient for all the data.
(ii) What can you say about the correlation?

A	74	82	78	72	79
B	34	42	37	39	45

(i) To answer this question, you must first rank each piece of data for A and then rank each piece of data for B. We will rank starting with the smallest piece of data. Therefore for A we would rank as follows:

    72 rank 1
    74 rank 2
    78 rank 3
    79 rank 4
    82 rank 5
And for rank B
    34 rank 1
    37 rank 2
    39 rank 3
    42 rank 4
    45 rank 5
This gives us the following table:

A	74	82	78	72	79
Rank A	2	5	3	1	4
B	34	42	37	39	45
Rank B	1	4	2	3	5

Next, you must work out the difference in the corresponding ranks. This gives:

Rank A	2	5	3	1	4
Rank B	1	4	2	3	5
Difference d	1	1	1	-2	-1

Next, you must square these differences. This gives:

d	1	1	1	-2	-1
d²	1	1	1	4	1

Then you must add these differences. This gives:

$$\Sigma d^2 = 1 + 1 + 1 + 4 + 1$$
$$= 8$$

You then calculate Spearman's Rank Correlation Coefficient, where n = 5:

$$r = 1 - \frac{6\Sigma d^2}{n(n^2 - 1)}$$

$$r = 1 - \frac{6 \times 8}{5 \times 24}$$

$$r = 1 - \frac{48}{120}$$

$$r = 0.6$$

(ii) As 0.6 is positive this shows there is **positive** correlation.

26.3 Tied Ranks in Spearman's Rank Correlation Coefficient

Where two or more of the data are the same then you need to use **tied ranks**. For example, when working with the following data:

A	50	45	50	53

we would rank starting with the smallest piece of data, ie 45 is rank 1. The first 50 should be rank 2, but it is tied with the other 50 which should be rank 3. In this case you take the mean of the ranks for both the 50s, ie:

    45 rank 1

    50 rank 2.5

    50 rank 2.5

    53 rank 4

Sometimes, more than two pieces of data are paired. For example, when working with this data:

A	50	45	50	50

we would again rank starting with the smallest piece of data, ie 45 is rank 1. The first 50 should be rank 2 but it is tied with the second 50 which should be rank 3 **and** the third 50 which should be rank 4. In this case you take the mean of the ranks for all three 50s, ie:

    45 rank 1

    50 rank 3

    50 rank 3

    50 rank 3

You can use this approach if you have to calculate Spearman's Rank Correlation Coefficient for tied data.

For example:

1. The table below shows pairs of data.

(i) Calculate Spearman's Rank Correlation Coefficient for all the data.

(ii) What can you say about the correlation?

A	5.6	4.2	9.3	6.4	4.2	3.8	7.5	5.2
B	54	52	22	35	54	61	32	36

(i) As before, you must first rank each piece of data for A and then rank each piece of data for B to get:

A	5.6	4.2	9.3	6.4	4.2	3.8	7.5	5.2
Rank A	5	2.5	8	6	2.5	1	7	4
B	54	52	22	35	54	61	32	36
Rank B	6.5	5	1	3	6.5	8	2	4

Next, you must work out the difference in the corresponding ranks. This gives:

Rank A	5	2.5	8	6	2.5	1	7	4
Rank B	6.5	5	1	3	6.5	8	2	4
Difference d	−1.5	−2.5	7	3	−4	−7	5	0

Next, you must square these differences. This gives:

d	−1.5	−2.5	7	3	−4	−7	5	0
d²	2.25	6.25	49	9	16	49	25	0

Then you must add these differences. This gives:
$$\Sigma d^2 = 156.5$$
You then calculate Spearman's Rank Correlation Coefficient, where n = 8:
$$r = 1 - \frac{6\Sigma d^2}{n(n^2 - 1)}$$
$$r = 1 - \frac{6 \times 156.5}{8 \times 63}$$
$$r = 1 - \frac{939}{504}$$
$$r = -0.863$$

(ii) As this value is less than 0 this shows there is **negative** correlation.

....

2. The table below shows pairs of data.

(i) Calculate Spearman's Rank Correlation Coefficient for all the data.

(ii) What can you say about the correlation?

A	43	45	34	41	50	37	39
B	99	99	92	87	99	84	95

(i) There are three 99's. These should be ranked 5, 6 and 7
So you rank each of them as 6, ie the mean of 5, 6 and 7:

A	43	45	34	41	50	37	39
Rank A	5	6	1	4	7	2	3
B	99	99	92	87	99	84	95
Rank B	6	6	3	2	6	1	4

And calculate the differences, and square them:

Rank A	5	6	1	4	7	2	3
Rank B	6	6	3	2	6	1	4
Difference d	−1	0	−2	2	1	1	−1
d²	1	0	4	4	1	1	1

Adding these differences gives:
$$\Sigma d^2 = 12$$
Then calculate Spearman's Rank Correlation Coefficient, where n = 7:
$$r = 1 - \frac{6\Sigma d^2}{n(n^2 - 1)}$$
$$r = 1 - \frac{6 \times 12}{7 \times 48}$$
$$r = 1 - \frac{72}{336}$$
$$r = 0.786$$

(ii) As 0.786 is positive, this shows there is **positive** correlation.

Exercise 26A: *For each of the following:*

(a) Calculate Spearman's Rank Correlation Coefficient for all the data.

(b) Comment on the correlation.

1.

A	61	56	52	64	56	58	43	48
B	68	65	68	71	74	62	58	54

2.

A	5.6	5.1	4.8	4.6	5.8	5.4	4.8	4.3
B	5.3	5.2	4.6	4.1	5.2	4.9	4.5	4.3

3.

A	63	48	52	73	84	69
B	42	46	47	38	38	48

4.

A	82	64	82	84	68	74	82	73
B	9	3	15	13	8	10	8	4

5.

A	4.6	3.6	3.4	4.8	4.2	4.3	3.4	3.8
B	44	74	69	52	57	44	62	42

6.

A	36	34	45	28	56	39	44	52
B	5.2	5.7	4.3	5.6	4.6	5.4	5	4.5

Exercise 26A...

7.

A	4.5	4.2	3.2	4.7	2.4	4.4	3.5	2.6
B	68	84	73	81	64	81	81	56

8.

A	58	69	61	56	67	58	64	72
B	52	43	52	55	48	59	57	45

9.

A	41	38	55	36	52	41	48	41
B	76	73	94	83	86	76	89	79

10.

A	25	14	28	11	22	26	19	14
B	8	18	4	16	2	3	6	17

26.4 Scatter Graphs

You draw a scatter graph by plotting the data for one variable on the horizontal axis and the data for the other variable on the vertical axis.

You can draw a **line of best fit** on the scatter graph if there is correlation. The rules for drawing a line of best fit are as follows:

• Work out the mean for each of the two sets of data
• Draw an envelope around the points plotted
• Plot the point showing the mean for each of the two sets of data
• Draw a straight line passing through this point with roughly the same number of points on each side and roughly parallel to the side of the envelope.

You can find the equation of the line of best fit by using these rules:

• Take 2 points that lie on the line of best fit (it is a good idea to use the point which is the mean for each of the two sets of data).
• Find the gradient, m, of the line of best fit by using:

$$m = \frac{y_2 - y_1}{x_2 - x_1}$$

Substitute the x-coordinate and the y-coordinate of one of the points into $y = mx + c$ to find the value of c.
For example:

The table below shows the marks of 6 pupils in English and Maths:

Maths	43	48	57	52	54	48
English	72	86	88	75	94	79

(a) Draw a scatter graph to show this data with the English marks on the vertical axis and the Maths marks on the horizontal axis.
(b) Draw the line of best fit.
(c) Determine the equation of the line of best fit.
(d) Calculate Spearman's Rank Correlation Coefficient.
(e) What significance do you attach to the value of Spearman's Rank Correlation Coefficient?

(a) You plot these points on graph paper (see below).
(b) You draw an envelope around the points that you plotted. You then find the mean English and the mean Maths scores:

English: Total = 494
            Mean = 494 ÷ 6 = 82.3
Maths: Total = 302
            Mean = 302 ÷ 6 = 50.3

You must plot these on the graph and draw the line of best fit through the point (50.3, 82.3):

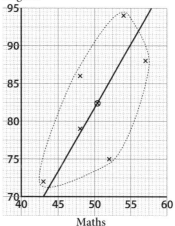

(c) You need to choose 2 points on the line of best fit in order to work out the gradient. Let us choose (50.3, 82.3) and (43, 70). Then:

$$m = \frac{y_2 - y_1}{x_2 - x_1}$$

$$m = \frac{82.3 - 70}{50.3 - 43}$$

$$m = \frac{12.3}{7.3}$$

$$m = 1.68$$

So $y = mx + c$
            $y = 1.68x + c$

You can now substitute one of these points into this equation to work out c. Let us choose (43, 70).

So $x = 43$ and $y = 70$ giving:
$$y = 1.68x + c$$
$$70 = 1.68 \times 43 + c$$
$$70 = 72.24 + c$$
So
$$c = -2.24$$
So the equation of the line of best fit is:
$$y = 1.68x - 2.24$$

(d)

English	72	86	88	75	94	79
Rank English	1	4	5	2	6	3
Maths	43	48	57	52	54	48
Rank Maths	1	2.5	6	4	5	2.5

Rank English	1	4	5	2	6	3
Rank Maths	1	2.5	6	4	5	2.5
Difference d	0	1.5	−1	−2	1	0.5

d	0	1.5	−1	−2	1	0.5
d^2	0	2.25	1	4	1	0.25

$$\Sigma d^2 = 8.5$$
$$n = 6$$
So
$$r = 1 - \frac{6\Sigma d^2}{n(n^2 - 1)}$$
$$r = 1 - \frac{6 \times 8.5}{6 \times 35}$$
$$r = 1 - \frac{51}{336}$$
$$r = 0.757$$

(e) As 0.757 is positive this shows there is **positive** correlation.

Exercise 26B:

1. The table below shows the marks of 5 pupils in History and French.

French	53	59	56	64	62
History	77	86	74	83	94

(a) Draw a scatter graph to show this data with the History marks on the vertical axis and the French marks on the horizontal axis.
(b) Draw the line of best fit.
(c) Determine the equation of the line of best fit.
(d) Calculate Spearman's Rank Correlation Coefficient.
(e) What significance do you attach to the value of Spearman's Rank Correlation Coefficient?

Exercise 26B...

2. The table below shows the weights and lengths of 7 objects.

Length cm	32	36	14	22	17	32	25
Weight kg	5.1	5.4	4.2	4.7	3.6	5.3	3.2

(a) Draw a scatter graph to show this data with the weights on the vertical axis and the lengths on the horizontal axis.
(b) Draw the line of best fit.
(c) Determine the equation of the line of best fit.
(d) Calculate Spearman's Rank Correlation Coefficient.
(e) What significance do you attach to the value of Spearman's Rank Correlation Coefficient?

3. The table below shows the area and volume of 8 objects.

Area cm^2	94	81	91	86	96	96	74	85
Volume cm^3	451	438	444	425	456	452	423	412

(a) Draw a scatter graph to show this data with the volume on the vertical axis and the area on the horizontal axis.
(b) Draw the line of best fit.
(c) Determine the equation of the line of best fit.
(d) Calculate Spearman's Rank Correlation Coefficient.
(e) What significance do you attach to the value of Spearman's Rank Correlation Coefficient?

4. The table below shows the density and volume of 7 objects.

Volume cm^3	115	97	115	111	123	94	104
Density kg/ cm^3	5.2	5.6	4.5	5.6	4.8	6.8	6.2

(a) Draw a scatter graph to show this data with the density on the vertical axis and the volume on the horizontal axis.
(b) Draw the line of best fit.
(c) Determine the equation of the line of best fit.
(d) Calculate Spearman's Rank Correlation Coefficient.
(e) What significance do you attach to the value of Spearman's Rank Correlation Coefficient?

5. The table below shows the perimeter and area of 7 objects.

Perimeter cm	52	61	55	58	64	48	46
Area cm^2	77	94	86	81	88	81	74

(a) Draw a scatter graph to show this data with the area on the vertical axis and the perimeter on the horizontal axis.

(b) Draw the line of best fit.
(c) Determine the equation of the line of best fit.
(d) Calculate Spearman's Rank Correlation Coefficient.
(e) What significance do you attach to the value of Spearman's Rank Correlation Coefficient?

6. The table below shows the mass and length of 9 objects.

Mass kg	3.5	12	19	25.5	31	28.5	24	7.5	15
Length cm	1.5	2.5	4.7	4.5	6.7	5.7	6.3	2.5	3.7

(a) Draw a scatter graph to show this data with the length on the vertical axis and the mass on the horizontal axis.
(b) Draw the line of best fit.
(c) Determine the equation of the line of best fit.
(d) Calculate Spearman's Rank Correlation Coefficient.
(e) What significance do you attach to the value of Spearman's Rank Correlation Coefficient?

7. The table below shows the average speed of a car and the time taken for 8 journeys.

Time taken hours	7.6	8.7	6.4	8.3	7.1	8.1	7.3	6.8
Average speed mph	40	36	52	32	62	45	48	56

(a) Draw a scatter graph to show this data with the average speed on the vertical axis and the time taken on the horizontal axis.
(b) Draw the line of best fit.
(c) Determine the equation of the line of best fit.
(d) Calculate Spearman's Rank Correlation Coefficient.
(e) What significance do you attach to the value of Spearman's Rank Correlation Coefficient?

8. The table below shows the temperature and rainfall at 6 different places.

Rainfall mm	12	26	13	21	21	9
Temperature °C	8	12	7	9	14	5

(a) Draw a scatter graph to show this data with the temperature on the vertical axis and the rainfall on the horizontal axis.
(b) Draw the line of best fit.
(c) Determine the equation of the line of best fit.
(d) Calculate Spearman's Rank Correlation Coefficient.
(e) What significance do you attach to the value of Spearman's Rank Correlation Coefficient?

9. The table below shows the area and width of 8 objects.

Width cm	31	40	50	24	51	43	32	47
Area cm²	31	46	76	24	71	61	41	58

(a) Draw a scatter graph to show this data with the area on the vertical axis and the width on the horizontal axis.
(b) Draw the line of best fit.
(c) Determine the equation of the line of best fit.
(d) Calculate Spearman's Rank Correlation Coefficient.
(e) What significance do you attach to the value of Spearman's Rank Correlation Coefficient?

10. The table below shows the perimeter and area of 7 objects.

Perimeter cm	43	45	34	41	50	37	39
Area cm²	99	97	92	87	99	84	95

(a) Draw a scatter graph to show this data with the area on the vertical axis and the perimeter on the horizontal axis.
(b) Draw the line of best fit.
(c) Determine the equation of the line of best fit.
(d) Calculate Spearman's Rank Correlation Coefficient.
(e) What significance do you attach to the value of Spearman's Rank Correlation?

Answers

Exercise 1A

1. $\dfrac{11x + 17}{12}$

2. $\dfrac{3x - 16}{10}$

3. $\dfrac{-1}{12}$

4. $\dfrac{5x - 12}{8}$

5. $\dfrac{11x - 8}{20}$

6. $\dfrac{-2x + 24}{(3x - 2)(x + 5)}$

7. $\dfrac{3x + 24}{(2 - x)(x + 4)}$

8. $\dfrac{8x - 18}{(3 - 2x)(x - 3)}$

9. $\dfrac{11x - 23}{(2x - 1)(x - 4)}$

10. $\dfrac{12x - 1}{(x - 3)(2x + 1)}$

Exercise 1B

1. $\dfrac{11x + 8}{x(x - 3)(x + 2)}$

2. $\dfrac{9x + 10}{2x(x - 3)(x + 2)}$

3. $\dfrac{7x + 5}{(x + 3)(x - 1)(x + 5)}$

4. $\dfrac{7x + 21}{(x + 4)(x - 2)(2x + 1)}$

5. $\dfrac{5x - 2}{x(x - 4)(x + 2)}$

6. $\dfrac{x + 26}{(x + 1)(x + 5)(x - 2)}$

7. $\dfrac{x + 20}{2x(x - 3)(x + 4)}$

8. $\dfrac{-x - 2}{(x + 4)(x - 1)(x - 2)}$

9. $\dfrac{-x + 10}{(x + 5)(x - 2)(x + 2)}$

10. $\dfrac{x - 16}{(x + 2)(x - 4)(2x - 1)}$

Exercise 1C

1. $\dfrac{13x + 4}{3x(x + 1)(x + 2)}$

2. $\dfrac{7x + 18}{(x + 2)(x - 2)(x + 4)}$

3. $\dfrac{13x - 14}{(x + 4)(x - 3)(2x - 1)}$

4. $\dfrac{10x - 3}{2x(x - 5)(2x - 1)}$

5. $\dfrac{7x - 5}{(x - 2)(x - 5)(x + 5)}$

6. $\dfrac{6}{x(x + 1)(2x - 1)}$

7. $\dfrac{x + 8}{(x + 2)(x - 4)(x + 4)}$

8. $\dfrac{-4x - 9}{x(x - 2)(2x + 3)}$

9. $\dfrac{3x + 35}{(x + 4)(x - 3)(2x + 5)}$

10. $\dfrac{-5x - 7}{(x + 2)(x - 3)(4x - 1)}$

Exercise 1D

1. $\dfrac{5xy}{v^2}$

2. $\dfrac{t}{nv}$

3. $\dfrac{4q^2}{rn}$

4. $\dfrac{9wt}{v^2}$

5. $\dfrac{10rw}{vq}$

6. $\dfrac{6a^3b^2}{5c}$

7. $\dfrac{8qn}{5}$

8. $\dfrac{2vt}{3wn}$

9. $\dfrac{18c}{7ab}$

10. $\dfrac{10qv^3}{3r}$

Exercise 1E

1. $\dfrac{2x - 1}{x - 2}$

2. $\dfrac{3x}{2x - 1}$

3. $\dfrac{x + 5}{x + 4}$

4. $\dfrac{4x - 3}{x - 3}$

5. $\dfrac{x + 5}{x + 2}$

6. $\dfrac{4x}{x + 2}$

7. $\dfrac{2x - 1}{x - 2}$

8. $\dfrac{x + 5}{2x + 3}$

9. $\dfrac{4x}{2x + 1}$

10. $\dfrac{5x - 2}{x - 5}$

Exercise 1F

1. $\dfrac{2x}{x + 4}$

2. $\dfrac{5(x - 3)}{2(3x - 2)}$

3. $\dfrac{3(x + 5)}{4(x - 2)}$

4. $\dfrac{3(3x + 2)}{2(x - 2)}$

5. $\dfrac{8(2x - 5)}{15(x + 4)}$

6. $\dfrac{8(x + 2)}{3x - 1}$

7. $\dfrac{x-7}{2(x+5)}$

8. $\dfrac{3(x+5)}{4(4x+3)}$

9. $\dfrac{4(x+6)}{3(x-2)}$

10. $\dfrac{3(4x-1)}{2(x-2)}$

Exercise 2A
1. $\frac{4}{3}$ or $-\frac{4}{3}$
2. $-\frac{2}{3}$ or 1
3. 0 or $-\frac{7}{2}$
4. 5 or $-\frac{3}{2}$
5. $-\frac{1}{2}$ or $\frac{2}{3}$
6. $-\frac{2}{3}$ or 4
7. -5 or $\frac{3}{4}$
8. 0 or $\frac{7}{3}$
9. $\frac{2}{3}$ or $-\frac{9}{2}$
10. 7 or $\frac{3}{2}$
11. $-\frac{1}{2}$ or $-\frac{2}{3}$
12. $\frac{2}{5}$ or $-\frac{2}{5}$

Exercise 2B
1. 5.45 or 0.55
2. 1.64 or -2.14
3. 2.10 or -1.43
4. -0.23 or -8.77
5. 2.14 or -1.64
6. 0.55 or -1.22
7. 1.37 or -1.17
8. -0.27 or -3.73
9. 0.44 or -1.69
10. 4.16 or -2.16

Exercise 2C
1. 4 or $\frac{7}{3}$
2. 2 or $\frac{2}{3}$
3. 1 or -6
4. 3 or $\frac{5}{3}$
5. 2 or $-\frac{1}{2}$
6. 5 or $\frac{5}{3}$
7. 3 or $\frac{5}{8}$
8. 2 or $-\frac{1}{2}$
9. 3 or $\frac{3}{2}$
10. 5 or $\frac{5}{2}$

Exercise 2D
1. 3 or $-\frac{5}{2}$
2. -3

3. 5 or $\frac{3}{2}$
4. 4 or $\frac{5}{4}$
5. -2 or 0
6. 1.90 or 0.88
7. 3 or $\frac{1}{2}$
8. -4 or 5
9. 3 or $\frac{2}{3}$
10. 6 or $\frac{10}{3}$

Exercise 2E
1. $-3 \pm\sqrt{12}$
2. $1 \pm\sqrt{6}$
3. $\dfrac{-1\pm\sqrt{17}}{2}$
4. $\dfrac{3\pm\sqrt{29}}{2}$
5. $-4 \pm\sqrt{11}$
6. $5 \pm\sqrt{27}$
7. $\dfrac{-5\pm\sqrt{41}}{2}$
8. $\dfrac{7\pm\sqrt{69}}{2}$
9. $-2 \pm\sqrt{13}$
10. $5 \pm\sqrt{28}$

Exercise 2F
1. (i) -12 (ii) -3
2. (i) -3 (ii) 2
3. (i) -11 (ii) -4
4. (i) -9 (ii) 1
5. (i) $-3\frac{1}{4}$ (ii) $\frac{1}{2}$
6. (i) $5\frac{3}{4}$ (ii) $-1\frac{1}{2}$
7. (i) 2 (ii) 3
8. (i) 10 (ii) -5

Exercise 3A
1. 5, 1, 4
2. 3, 2, 5
3. 4, 2, 3
4. 2, -5, 3
5. 4, -1, -2
6. -2, 3, 4
7. 5, 2, 4
8. 7, -3, 5
9. 2, 4, -5
10. 5, 3, 6

Exercise 3B
1. (iv) 60p, 35p, 20p (v) £39.10

2. (iv) 8, 5, 3 (v) 112
3. (iv) £160, £120, £85 (v) £170
4. (iv) £20, £30, £90 (v) 35%
5. (iv) £6.50, £4.75, £5.50 (v) 12
6. (iv) 600g, 950g, 870g (v) 21, 24, 15
7. 66

Exercise 3C
1. $x = 4, y = 7$ and $x = -\frac{8}{3}, y = -\frac{19}{3}$
2. $x = 5, y = 2$ and $x = -7, y = -1$
3. $x = 4, y = -3$ and $x = -\frac{2}{5}, y = \frac{18}{5}$
4. $x = -2, y = 6$ and $x = -8, y = -6$
5. $x = 6, y = 3$ and $x = -2\frac{1}{4}, y = -\frac{3}{4}$
6. $x = 5, y = 4$ and $x = -5, y = -1$
7. $(7, -2)$ and $(-\frac{19}{2}, \frac{7}{2})$
8. $(-5, 3)$ and $(1, 9)$
9. $(2, -1)$ and $(-1, -2)$
10. $(-2, 4)$ and $(\frac{14}{3}, -6)$

Exercise 4A
1. 62.5° or 297.5°
2. 229.6° or 310.4°
3. 71.6° or 251.6°
4. 11.5° or 168.5°
5. 138.6° or 221.4°
6. 149.04° or 329.04°

Exercise 4B
1. 53.1° or -53.1°
2. -19.9° or -160.1°
3. 116.6° or -63.4°
4. 47.6° or 132.4°
5. 110.4° or -110.4°
6. 123.7° or -56.3°

Exercise 4C
1. 9.28°, 50.7°
2. 49.6°, 130.4°
3. 13.1°, 58.1°
4. 213.7°, 241.3°
5. 108.1°, 301.9°
6. -16.6°, 163.4°
7. -108.3°, 390.3°
8. 15°, 75°
9. 37.2°, -30.5°
10. 88.4°, -91.6°
11. 27.8°, -8.16°
12. 350°, 670°

Exercise 5A
1. 9.84 cm

2. 27.4°
3. 1.72 cm
4. 11.5 cm
5. 51.3°
6. 11.4 cm
7. 38.6 cm
8. 6.43 cm

Exercise 5B
1. 50.4°
2. 63.7°
3. 8.55 cm
4. 4.83 cm
5. 51.7°
6. 5.76 cm
7. 6.27 cm
8. 94.9°

Exercise 5C
1. 26.7 cm^2
2. 26.6 cm^2
3. 5.03 cm^2
4. 35.9 cm^2
5. 28.3 cm^2
6. 6.68 cm^2
7. 2.92 cm^2
8. 11.97 cm^2

Exercise 5D
1(i). 5.88 cm (ii) 30.7° (iii) 48.3 cm^2
2(i). 52.7° (ii) 5.24 cm (iii) 4.75 cm
3(i). 9.64 cm (ii) 114° (iii) 70.6 cm^2
4(i). 18 km (ii) 172.8°
5(i). 8.26 cm (ii) 116.6° (iii) 31.92 cm

Exercise 6A
1. 6
2. $4x - 3$
3. $6x + 1$
4. $6x^2 + 6x - 5$
5. $12x^2 - 12x + 1$
6. $3 - 2x + 21x^2$
7. $-3 + 4x - 3x^2$
8. $4x^3 - 4x$
9. $18x^2 - 3$
10. $8x - 3x^2 + 8x^3$
11. $10x^4 - 12x^3 + 6x^2 - 12x + 1$
12. $12x^2 - 14x$
13. $24x^3 - 4x + 1$
14. $3 - 3x^2$
15. $6x^2 - 8x - 7$

16. $-2 + 6x - 3x^2$
17. $6x^2 - 2x + 4$
18. $8x - 5$
19. $2ax + b$
20. $24x^2 - v$
21. $18x + t$

Exercise 6B
1. $\frac{3}{4}x^2 - \frac{4}{5}x + 1$
2. $6x^2 - \frac{2}{5}x + \frac{3}{4}$
3. $12x^2 - \frac{4}{7}x + \frac{1}{3}$
4. $\frac{3}{5}x^2 - 3x + 4$
5. $\frac{2}{3}x^2 + \frac{1}{2}x$
6. $\frac{3}{2}x + \frac{1}{5}$
7. $9x^2 - \frac{3}{2}x + 1$
8. $-\frac{1}{5} + 5x$
9. $7x^2 - \frac{4}{5}x + \frac{1}{4}$
10. $\frac{1}{2} - \frac{5}{3}x + 7x^2$

Exercise 6C
1. $\frac{-12}{x^3}$
2. $\frac{-12}{x^4}$
3. $\frac{-20}{x^5}$
4. $\frac{-21}{2x^8}$
5. $\frac{-5}{2x^3}$
6. $\frac{-6}{7x^4}$
7. $\frac{-12}{x^4}$
8. $\frac{-15}{x^6}$
9. $\frac{-3}{2x^4}$
10. $\frac{-1}{x^3}$

Exercise 6D
1. $18x^2 - \frac{8}{x^5}$
2. $2 - \frac{-15}{x^6}$
3. $10x^4 + 3 + \frac{8}{x^3}$
4. $4x^5 + \frac{4x}{3}$
5. $3x^3 + 1 + \frac{9}{4x^4}$
6. $7x + \frac{4}{7x^3}$
7. $12x^2 + 2x + \frac{4}{x^5}$

8. $4x - \frac{2}{3x^3}$
9. $8x - 7 - \frac{4}{x^3}$
10. $5 - \frac{8}{3x^5}$
11. $14x - \frac{12}{x^7}$
12. $12x^3 - \frac{10}{x^6} - 3$
13. $8x^5 + \frac{9}{2x^7}$
14. $8x^3 + 5 + \frac{10}{3x^6}$
15. $15x^4 + 2x^2 + \frac{3}{x^3}$

Exercise 6E
1. $12x^3 - 21x^2 + 4x$, $36x^2 - 42x + 4$
2. $6 + 4x$, 4
3. $6x^2 + 14x - 5$, $12x + 14$
4. $3 - \frac{4}{x^3}$, $\frac{12}{x^4}$
5. $-1 - 2x$, -2
6. $4 - 4x$, -4
7. $12x - \frac{2}{x^3}$, $12 + \frac{6}{x^4}$
8. $2x^2 - 4$, $4x$
9. $6x^9 - \frac{2}{5x^3}$, $54x^8 + \frac{6}{5x^4}$
10. $3x - \frac{4}{3x^3}$, $3 + \frac{4}{x^4}$
11. $12x^2 + \frac{6}{x^4}$, $24x - \frac{24}{x^5}$
12. $12x + \frac{1}{2x^4}$, $12 - \frac{2}{x^5}$
13. $6x - 1 + \frac{2}{x^3}$, $6 - \frac{6}{x^4}$
14. $2 + 6x + \frac{12}{x^4}$, $6 - \frac{48}{x^5}$
15. $6 - \frac{8}{5x^5}$, $\frac{8}{x^6}$

Exercise 7A
1. 1
2. 27
3. −5
4. 10
5. $3 \frac{8}{15}$
6. 2
7. 16
8. −1
9. $-\frac{3}{16}$
10. $6 \frac{1}{4}$
11. $-1 \frac{1}{4}$
12. 17
13. 25
14. $25 \frac{1}{4}$
15. $\frac{3}{8}$

Exercise 7B

1. (3, 6)
2. (−2, −2)
3. (4, 47)
4. (−1, 13)
5. (2, 1)
6. (−2, −2)
7. (−3, −7)

8. (3, 16)
9. (−4, 64)
10. (2, 6)
11. ($\frac{5}{8}$, 5 $\frac{7}{16}$)
12. (−$\frac{2}{5}$, −1 $\frac{2}{5}$)
13. (−$\frac{1}{3}$, 7 $\frac{1}{3}$)

Exercise 7C

1. (1, 1) and (0, −2)
2. (2, 37) and (−$\frac{11}{6}$, −26 $\frac{73}{108}$)
3. (−1, −9) and ($\frac{13}{9}$, 10 $\frac{157}{243}$)
4. (−2, 11) and ($\frac{5}{3}$, −5 $\frac{19}{27}$)
5. (2, 8) and (−$\frac{2}{3}$, −6 $\frac{14}{27}$)
6. (1, −$\frac{1}{6}$) and (−4, 15 $\frac{2}{3}$)
7. (−1, −9) and ($\frac{5}{3}$, 5 $\frac{14}{27}$)
8. (2, −2 $\frac{2}{3}$) and (−$\frac{1}{2}$, −4 $\frac{23}{24}$)
9. (−1, −10) and ($\frac{4}{3}$, 13 $\frac{16}{27}$)
10. (2, 7) and (0, 7)
11. ($\frac{1}{2}$, −2 $\frac{3}{4}$) and (−3, 83)
12. (− $\frac{2}{3}$, −$\frac{5}{27}$) and (4, −51)
13. (4, −75) and (−2, 33)

Exercise 7D

1. $y = 7x − 14$
2. $y = 12x − 8$
3. $y = −11x + 31$
4. $y = 23x + 30$
5. $y = 19x + 9$
6. $y = 25x + 20$
7. $y = 4x + 6$
8. $y = 4x − 5 \frac{1}{2}$
9. $y = − \frac{1}{4}x + 2$
10. $y = \frac{3}{2}x + 1$

Exercise 7E

1. $y = −\frac{1}{17}x + 14 \frac{2}{17}$
2. $y = −\frac{1}{7}x + \frac{1}{7}$
3. $y = −\frac{1}{21}x − 23 \frac{2}{21}$
4. $y = −\frac{1}{19}x − 11 \frac{1}{9}$
5. $y = \frac{1}{2}x + 6$
6. $y = \frac{1}{11}x + 7 \frac{10}{11}$
7. $y = \frac{4}{3}x − \frac{23}{12}$
8. $y = −\frac{1}{6}x − 5$
9. $y = 2x + 3$
10. $y = \frac{4}{31}x + 4 \frac{47}{124}$

Exercise 7F

1. (a) $y = 13x−11$ (b) (i) (0, −11)
   (ii) ($\frac{11}{13}$, 0)

2. (a) $y = \frac{1}{9}x + 13 \frac{1}{3}$
   (b) (i) (0, 13 $\frac{1}{3}$) (ii) (−120, 0)
3. (a) $y = −6x + 3$ (b) (−3, 21)
4. (a) $y = \frac{1}{9}x + \frac{7}{9}$ (b) (5, $\frac{4}{3}$)
5. (a) $y = 4x + \frac{11}{6}$ (b) (0, $\frac{11}{6}$)
6. (a) $y = \frac{1}{6}x + 1 \frac{1}{6}$ (b) (i) (−7, 0)
   (ii) (2, 1 $\frac{1}{2}$)
7. (a) $y = 11x + 25$ (b) (i) (−$\frac{25}{11}$, 0)
   (ii) (−3, −8)
8. (a) $y = −\frac{1}{7}x − 2 \frac{1}{7}$ (b) (i) (0, −2 $\frac{1}{7}$)
   (ii) (6, −3)
9. (a) $y = −x − 4$ (b) (i) (0, −4)
   (ii) (5, −9)
10. (a) $y = −\frac{1}{5}x + 2 \frac{3}{10}$
    (b) (i) (11 $\frac{1}{2}$, 0) (ii) (4, 1 $\frac{1}{2}$)

Exercise 7G

1. (2, 4)
2. (1, 6)
3. (1, −4) ($\frac{1}{3}$, −3 $\frac{23}{27}$)
4. (1, 5 $\frac{1}{2}$) (−$\frac{1}{6}$, 7 $\frac{89}{216}$)
5. ($\frac{4}{3}$, 3) (−$\frac{4}{3}$, −3)
6. (4, −4)
7. (1 $\frac{1}{2}$, −8 $\frac{1}{4}$)
8. ($\frac{2}{3}$, 8 $\frac{2}{3}$) (−$\frac{2}{3}$, −8 $\frac{2}{3}$)
9. (2, −5) ($\frac{2}{3}$, −1 $\frac{4}{27}$)
10. (5, −43 $\frac{1}{6}$) (−$\frac{1}{2}$, 1 $\frac{7}{24}$)

Exercise 8A

1. (2, −11) Min
2. (−2, −11) Min
3. (3, 18) Max
4. (2 $\frac{1}{2}$, −9 $\frac{1}{4}$) Min
5. (−2, 11) Max
6. (3 $\frac{1}{2}$, 13 $\frac{1}{4}$) Max
7. (−$\frac{1}{4}$, −5 $\frac{1}{4}$) Min
8. (−$\frac{1}{2}$, 5 $\frac{1}{4}$) Max
9. (3, −16) Min   10. (−2, 12) Max

Exercise 8B

1. (2, −40) Min (−3, 85) Max
2. (4, −109) Min (−1, 16) Max
3. (4, 9) Min (2, 13) Max
4. ($\frac{1}{3}$, −2 $\frac{14}{27}$) Min (−3, 16) Max
5. ($\frac{1}{2}$, 1) Min (−$\frac{1}{2}$, 3) Max
6. (2, 6) Min ($\frac{4}{3}$, 6 $\frac{4}{27}$) Max
7. ($\frac{5}{6}$, 6 $\frac{37}{108}$) Max (3, −14) Min
8. ($\frac{2}{3}$, −2 $\frac{22}{27}$) Min (−4, 48) Max
9. ($\frac{5}{3}$, −31 $\frac{7}{27}$) Min (−$\frac{3}{2}$, 32 $\frac{1}{4}$) Max
10. (4, −184) Min (−6, 316) Max
11. (−5, −170) Min (3, 86) Max
12. (−1, 8) Max (−2, 7) Min

13. ($\frac{1}{2}$, 21 $\frac{1}{4}$) Min (2, 28) Max

Exercise 8C

1.
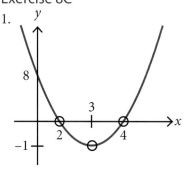

2.
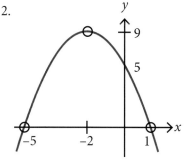

3.
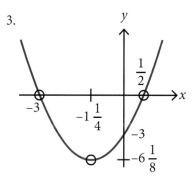

4.
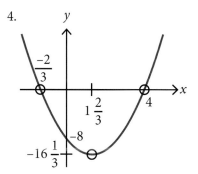

5.
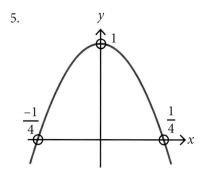

6.

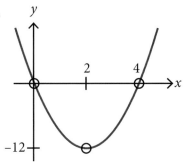

7.

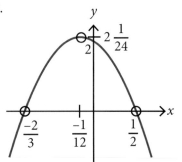

8.

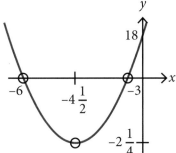

9.

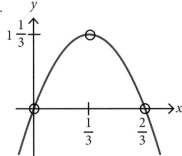

10.

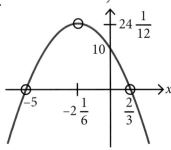

Exercise 8D

1.

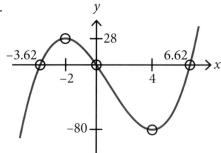

2.

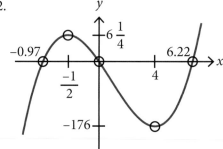

3.

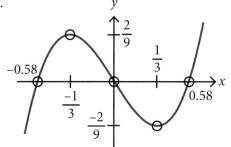

4.

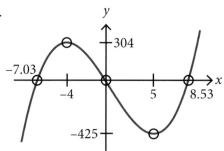

5.

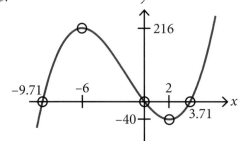

6.

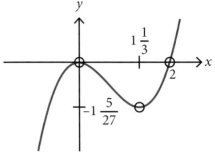

7.

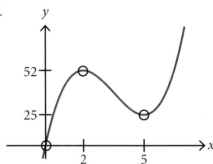

8.

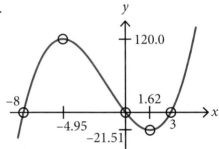

9.

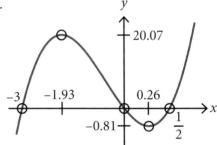

10.

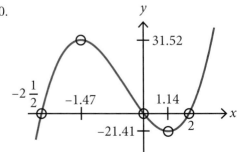

Exercise 8E

1. –768
2. (i) $x = 30$ and $y = 15$ (ii) 450 m²
3. (i) $x = 5$ and $y = 15$ (ii) $^{75}/_2$ cm²
4. (i) $x = 10$ and $y = 10$ (ii) £3
5. (i) $x = 24$ and $y = 4$ (ii) 48 m
6. (i) $x = 26$ and $y = 13$ (ii) 52 m
7. (i) $x = 2$ and $y = 4$ (ii) 64 cm²
8. (i) $x = 3$ and $y = 2$ (ii) 24 cm

Exercise 9A

1. $3x^3 + 3x + C$
2. $4x^3 + 2x^2 – 2x + C$
3. $4x^4 – 3x^3 – 3x^2 + 3x + C$
4. $7x – x^2 + C$
5. $2x^2 + x^3 + C$
6. $^3/_2x^4 – 2x^2 – x + C$
7. $^3/_5x^5 – ^1/_2x^4 + ^7/_3x^3 + C$
8. $^2/_{25}x^5 + 3x^2 + C$
9. $^1/_7x^3 – x^2 + C$
10. $^5/_{32}x^4 + ^1/_{12}x^3 + ^1/_2x^2 + C$
11. $3x^2 – ^4/_3 x^3 + 7x + C$
12. $9x – x^2 – 2x^3 + C$
13. $5x^2 + ^1/_3x^3 – 3x + C$
14. $^1/_6x^4 + ^2/_5x^3 – x + C$
15. $2x^2 – ^2/_{25}x^5 + 5x + C$
16. $2x^2 – 7x + C$
17. $x^4 – ^7/_3x^3 + ^5/_2x^2 – 2x + C$
18. $3x^2 – ^1/_9x^6 + C$

Exercise 9B

1. $\dfrac{-4}{x} + C$
2. $\dfrac{-4}{x^2} + C$
3. $\dfrac{-1}{2x^4} + C$
4. $\dfrac{-3}{5x} + C$
5. $\dfrac{-1}{7x^2} + C$
6. $\dfrac{-1}{5x^3} + C$
7. $2x^2 + \dfrac{4}{x} + C$
8. $\dfrac{x^3}{3} + 3x + \dfrac{1}{x^2} + C$
9. $\dfrac{x^3}{2} + \dfrac{2}{3x} + C$
10. $6x – \dfrac{1}{2x} + C$
11. $2x^2 + \dfrac{1}{8x^4} + C$
12. $\dfrac{-5}{3x} + \dfrac{x^3}{3} + C$
13. $\dfrac{x^4}{2} + \dfrac{1}{4x^2} + C$
14. $3x^2 + 2x + \dfrac{4}{x} + C$
15. $x^3 – 5x – \dfrac{2}{5x} + C$

Exercise 9C

1. 3
2. 3
3. $^{15}/_{32}$
4. $^2/_5$
5. $–^3/_{28}$

6. $^7/_{40}$

7. 21 ⅓

8. 24 $^{23}/_{48}$

9. 3 ⅙

10. 12 ⅓

11. −5 $^{111}/_{128}$

12. 10 $^5/_{18}$

13. 0

14. −33

15. 9 $^5/_6$

Exercise 9D

1. (i) $y = 3x^2 - 3x + 3$ (ii) 39

2. (i) $y = 4x + x^2 + 5$ (ii) 17

3. (i) $y = x^3 - 2x^2 - 3$ (ii) −48

4. (i) $y = 3x^2 + \frac{1}{3}x^3 - 5\frac{2}{3}$ (ii) 30 ⅓

5. (i) $y = \dfrac{-4}{x} + 6$ (ii) (−2, 8)

6. (i) $y = 2x^3 - x^2 + 5x - 59$ (ii) (4, 73)

7. (i) $y = \frac{3}{2}x^2 + 4x - 3\frac{1}{2}$ (ii) (−2, −5 ½)

8. (i) $y = 6x - x^2 + x^3 + 28$ (ii) (6, 244)

Exercise 10A

1. 8 ⅔   2. 12 ⅔

3. 16 ⅔   4. 42

5. 18 ⅔   6. 40 ⅔

7. 5 ⅙   8. ⅔

9. 11   10. 8

11. 55 ½   12. 22 ⅔

13. 9 ⅓   14. 27 $^5/_6$

15. 12

Exercise 10B

1. (i) 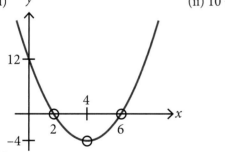 (ii) 10 ⅔

2. (i) 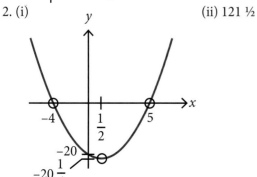 (ii) 121 ½

3. (i) 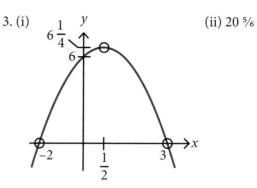 (ii) 20 $^5/_6$

4. (i) 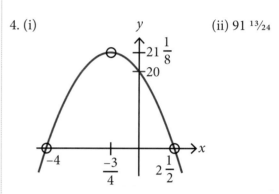 (ii) 91 $^{13}/_{24}$

5. (i) 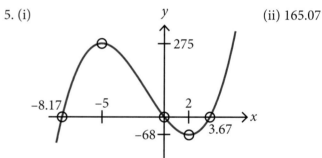 (ii) 165.07

6. (i) 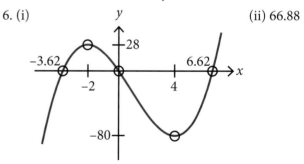 (ii) 66.88

7. (i) 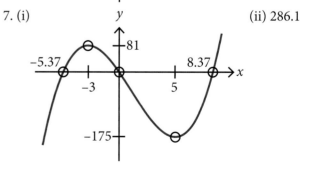 (ii) 286.1

8. (i)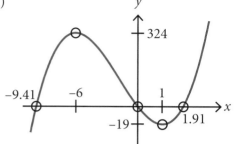

(ii) 24.2

9. (i)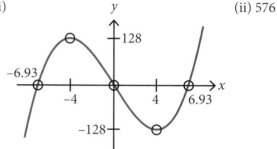

(ii) 576

10. (i)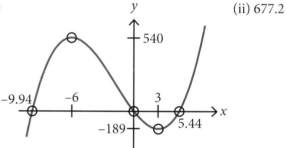

(ii) 677.2

11. (i)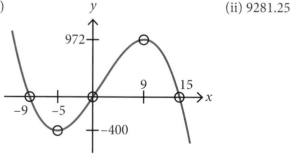

(ii) 9281.25

Exercise 11A

1. $\begin{pmatrix} -1 \\ 3 \end{pmatrix}$

2. $\begin{pmatrix} -4 \\ -4 \end{pmatrix}$

3. $\begin{pmatrix} 4 & 2 \end{pmatrix}$

4. $\begin{pmatrix} 3 & -1 \end{pmatrix}$

5. $\begin{pmatrix} -x \\ -2 \end{pmatrix}$

6. $\begin{pmatrix} 2 & -4y \end{pmatrix}$

7. $\begin{pmatrix} -x \\ -y \end{pmatrix}$

8. $\begin{pmatrix} -4x & 3y \end{pmatrix}$

9. $\begin{pmatrix} -5 & -10 \\ 14 & -2 \end{pmatrix}$

10. $\begin{pmatrix} 0 & -15 \\ -1 & 11 \end{pmatrix}$

11. $\begin{pmatrix} 4x & 3y \\ 6 & 2t \end{pmatrix}$

12. $\begin{pmatrix} -1 & 2 \\ -7 & 6 \end{pmatrix}$

13. $\begin{pmatrix} 10n & -3m \\ -14 & -3q \end{pmatrix}$

14. $\begin{pmatrix} -2.8 & 2.2 \\ -7.5 & 7.5 \end{pmatrix}$

15. $\begin{pmatrix} 4.9 & 8.9 \\ 1.7 & -11.8 \end{pmatrix}$

16. $\begin{pmatrix} -8 & -5 \\ 9 & 3 \end{pmatrix}$

17. (i) $\begin{pmatrix} -2 & 15 \\ -5 & 10 \end{pmatrix}$ (ii) $\begin{pmatrix} 10 & -1 \\ 1 & 2 \end{pmatrix}$ (iii) $\begin{pmatrix} -8 & -1 \\ 5 & 10 \end{pmatrix}$

(iv) $\begin{pmatrix} 2 & 0 \\ -3 & -6 \end{pmatrix}$ (v) $\begin{pmatrix} 2 & -2 \\ 6 & 12 \end{pmatrix}$ (vi) $\begin{pmatrix} -2 & 17 \\ -14 & -8 \end{pmatrix}$

(vii) $\begin{pmatrix} -10 & 16 \\ -9 & 2 \end{pmatrix}$ (viii) $\begin{pmatrix} -6 & -16 \\ 10 & 0 \end{pmatrix}$

Exercise 11B

1. $\begin{pmatrix} 25 & 30 \\ -10 & 20 \end{pmatrix}$

2. $\begin{pmatrix} -9 & -6 \\ 18 & -15 \end{pmatrix}$

3. $\begin{pmatrix} -5 & 20 \\ -15 & 30 \end{pmatrix}$

4. $\begin{pmatrix} 24 & -40 \\ -40 & 24 \end{pmatrix}$

5. $\begin{pmatrix} -63 & 56 \\ 21 & -42 \end{pmatrix}$

6. $\begin{pmatrix} 2½ & 3 \\ -1 & 2 \end{pmatrix}$

7. $\begin{pmatrix} 0.3 & -0.5 \\ -0.5 & 0.3 \end{pmatrix}$

8. $\begin{pmatrix} -1½ & -1 \\ 3 & -2½ \end{pmatrix}$

9. $\begin{pmatrix} 7 & 10 \\ 2 & 3 \end{pmatrix}$

10. $\begin{pmatrix} 0 & -7 \\ 14 & -21 \end{pmatrix}$

11. $\begin{pmatrix} -14 & 0 \\ 18 & -36 \end{pmatrix}$

12. $\begin{pmatrix} -30 & -28 \\ 36 & -37 \end{pmatrix}$

13. $\begin{pmatrix} 22 & -1 \\ -12 & 17 \end{pmatrix}$

14. $\begin{pmatrix} -24 & 38 \\ 1 & -2 \end{pmatrix}$

15. $\begin{pmatrix} -2 & 13 \\ 6 & 2 \end{pmatrix}$

16. (i) $\begin{pmatrix} -6 \\ 15 \end{pmatrix}$ (ii) $\begin{pmatrix} 6 & -8 \end{pmatrix}$ (iii) $\begin{pmatrix} 8 \\ -28 \end{pmatrix}$ (iv) $\begin{pmatrix} 2 & 10 \end{pmatrix}$

(v) $\begin{pmatrix} -1 \\ 2½ \end{pmatrix}$ (vi) $\begin{pmatrix} 2¼ & -3 \end{pmatrix}$ (vii) $\begin{pmatrix} -12 \\ 36 \end{pmatrix}$ (viii) $\begin{pmatrix} 1 & -4½ \end{pmatrix}$

Exercise 11C

1. $\begin{pmatrix} 2 & 5 \\ 10 & 12 \end{pmatrix}$

2. $\begin{pmatrix} 9 & -7 \\ 5 & -2 \end{pmatrix}$

3. $(16 \ -6 \ 1)$

4. $\begin{pmatrix} -3 \\ -4 \end{pmatrix}$

5. $\begin{pmatrix} 5 & 1 \\ 2 & -2 \end{pmatrix}$

6. $\begin{pmatrix} 3 & -2 \\ 1 & -2 \end{pmatrix}$

7. $\begin{pmatrix} 1 & -1 \\ -1 & 3 \end{pmatrix}$

8. $\begin{pmatrix} 2 & 3 \\ -6 & 7 \end{pmatrix}$

9. $\begin{pmatrix} 2 & -1 \\ -3 & -1 \end{pmatrix}$

10. $\begin{pmatrix} 3 & 3 \\ -4 & 3 \end{pmatrix}$

Exercise 11D

1. (-22)

2. $\begin{pmatrix} -8 & 16 \\ -4 & 8 \end{pmatrix}$

3. $\begin{pmatrix} 14 \\ 13 \end{pmatrix}$

4. (-1)

5. $\begin{pmatrix} -10 & 6 \\ 15 & -9 \end{pmatrix}$

6. $\begin{pmatrix} -14 \\ 4 \end{pmatrix}$

7. (-16)

8. $\begin{pmatrix} -12 & 6 \\ 20 & -10 \end{pmatrix}$

9. $\begin{pmatrix} -20 \\ 30 \end{pmatrix}$

10. (i) $\begin{pmatrix} -17 & 29 \\ -5 & 7 \end{pmatrix}$ (ii) $\begin{pmatrix} 19 & -30 \\ 1 & 8 \end{pmatrix}$ (iii) $\begin{pmatrix} -8 & 13 \\ 21 & -35 \end{pmatrix}$

(iv) $\begin{pmatrix} -10 & -1 \\ 26 & 0 \end{pmatrix}$ (v) $\begin{pmatrix} 24 & 5 \\ -22 & 3 \end{pmatrix}$ (vi) $\begin{pmatrix} -17 & 29 \\ 15 & -26 \end{pmatrix}$

(vii) $\begin{pmatrix} 24 & 5 \\ 4 & 29 \end{pmatrix}$ (viii) $\begin{pmatrix} 7 & -12 \\ -18 & 31 \end{pmatrix}$ (ix) $\begin{pmatrix} 19 & -30 \\ -18 & 31 \end{pmatrix}$

Exercise 11E

1. -18
2. 58
3. -3
4. -44
5. 72
6. -7
7. 17

8. -22
9. 18
10. -16

Exercise 11F

1. $\frac{1}{48} \begin{pmatrix} 2 & 6 \\ -7 & 3 \end{pmatrix}$

2. $\frac{-1}{2} \begin{pmatrix} -4 & 2 \\ -5 & 3 \end{pmatrix}$ or $\frac{1}{2} \begin{pmatrix} 4 & -2 \\ 5 & -3 \end{pmatrix}$

3. $\frac{-1}{24} \begin{pmatrix} 2 & -4 \\ -5 & -2 \end{pmatrix}$ or $\frac{1}{24} \begin{pmatrix} -2 & 4 \\ 5 & 2 \end{pmatrix}$

4. $\frac{-1}{14} \begin{pmatrix} -2 & -5 \\ -4 & -3 \end{pmatrix}$ or $\frac{1}{14} \begin{pmatrix} 2 & 5 \\ 4 & 3 \end{pmatrix}$

5. $\frac{-1}{7} \begin{pmatrix} -4 & -5 \\ -3 & -2 \end{pmatrix}$ or $\frac{1}{7} \begin{pmatrix} 4 & 5 \\ 3 & 2 \end{pmatrix}$

6. $\begin{pmatrix} -3 & -2 \\ -8 & -5 \end{pmatrix}$

7. $\frac{-1}{18} \begin{pmatrix} 3 & -6 \\ -4 & -2 \end{pmatrix}$ or $\frac{1}{18} \begin{pmatrix} -3 & 6 \\ 4 & 2 \end{pmatrix}$

8. $\frac{1}{19} \begin{pmatrix} 4 & 3 \\ -5 & 1 \end{pmatrix}$

9. $\frac{-1}{26} \begin{pmatrix} 2 & -4 \\ -5 & -3 \end{pmatrix}$ or $\frac{1}{26} \begin{pmatrix} -2 & 4 \\ 5 & 3 \end{pmatrix}$

10. $\begin{pmatrix} -5 & -3 \\ 3 & 2 \end{pmatrix}$

11. $\frac{-1}{2} \begin{pmatrix} -2 & -4 \\ -3 & -5 \end{pmatrix}$ or $\frac{1}{2} \begin{pmatrix} 2 & 4 \\ 3 & 5 \end{pmatrix}$

12. $\frac{1}{13} \begin{pmatrix} 3 & 2 \\ -5 & 1 \end{pmatrix}$

13. $\frac{1}{14} \begin{pmatrix} 3 & -4 \\ 5 & -2 \end{pmatrix}$

14. $\frac{-1}{33} \begin{pmatrix} -2 & -5 \\ -7 & -1 \end{pmatrix}$ or $\frac{1}{33} \begin{pmatrix} 2 & 5 \\ 7 & 1 \end{pmatrix}$

15. $\frac{1}{38} \begin{pmatrix} 4 & 2 \\ -7 & 6 \end{pmatrix}$

16. $\frac{-1}{14} \begin{pmatrix} -3 & -5 \\ -4 & -2 \end{pmatrix}$ or $\frac{1}{14} \begin{pmatrix} 3 & 5 \\ 4 & 2 \end{pmatrix}$

17. det = 0 You cannot divide 1 by 0. Therefore it has no inverse.

18. det = 0 You cannot divide 1 by 0. Therefore it has no inverse.

19. 6
20. $-{}^{20}\!/_3$

Exercise 11G

1. $\frac{-1}{14} \begin{pmatrix} -41 \\ 36 \end{pmatrix}$ or $\frac{1}{14} \begin{pmatrix} 41 \\ -36 \end{pmatrix}$

2. $\frac{-1}{6} \begin{pmatrix} -10 \\ -17 \end{pmatrix}$ or $\frac{1}{6} \begin{pmatrix} 10 \\ 17 \end{pmatrix}$

3. $\begin{pmatrix} 0 \\ -1 \end{pmatrix}$

4. $\frac{-1}{14} \begin{pmatrix} 18 \\ -10 \end{pmatrix}$ or $\frac{1}{14} \begin{pmatrix} -18 \\ 10 \end{pmatrix}$

5. $\frac{-1}{14} \begin{pmatrix} -13 & 14 \\ -6 & 0 \end{pmatrix}$ or $\frac{1}{14} \begin{pmatrix} 13 & -14 \\ 6 & 0 \end{pmatrix}$

6. $\dfrac{-1}{13}\begin{pmatrix} -36 & -41 \\ -14 & -21 \end{pmatrix}$ or $\dfrac{1}{13}\begin{pmatrix} 36 & 41 \\ 14 & 21 \end{pmatrix}$

7. $\dfrac{-1}{6}\begin{pmatrix} 14 & -36 \\ 22 & -51 \end{pmatrix}$ or $\dfrac{1}{6}\begin{pmatrix} -14 & 36 \\ -22 & 51 \end{pmatrix}$

8. $\dfrac{-1}{14}\begin{pmatrix} -21 & 41 \\ 14 & -36 \end{pmatrix}$ or $\dfrac{1}{14}\begin{pmatrix} 21 & -41 \\ -14 & 36 \end{pmatrix}$

9. $\dfrac{-1}{13}\begin{pmatrix} -51 & 36 \\ -22 & 14 \end{pmatrix}$ or $\dfrac{1}{13}\begin{pmatrix} 51 & -36 \\ 22 & -14 \end{pmatrix}$

10. $\dfrac{-1}{6}\begin{pmatrix} 0 & -14 \\ 6 & -13 \end{pmatrix}$ or $\dfrac{1}{6}\begin{pmatrix} 0 & 14 \\ -6 & 13 \end{pmatrix}$

11. $\dfrac{-1}{14}\begin{pmatrix} -8 & 27 \\ 20 & -36 \end{pmatrix}$ or $\dfrac{1}{14}\begin{pmatrix} 8 & -27 \\ -20 & 36 \end{pmatrix}$

12. $\dfrac{-1}{6}\begin{pmatrix} 26 \\ 34 \end{pmatrix}$ or $\dfrac{1}{6}\begin{pmatrix} -26 \\ -34 \end{pmatrix}$

13. $\dfrac{-1}{13}\begin{pmatrix} -225 & 26 \\ -94 & 0 \end{pmatrix}$ or $\dfrac{1}{13}\begin{pmatrix} 225 & -26 \\ 94 & 0 \end{pmatrix}$

14. $\dfrac{-1}{14}\begin{pmatrix} 159 \\ -128 \end{pmatrix}$ or $\dfrac{1}{14}\begin{pmatrix} -159 \\ 128 \end{pmatrix}$

Exercise 11H

1. $x = 4$ $\quad y = -2$
2. $x = -2$ $\quad y = 5$
3. $x = 3$ $\quad y = -2$
4. $x = 1\frac{1}{2}$ $\quad y = 4$
5. $x = -3$ $\quad y = -4$
6. $x = 2$ $\quad y = -4$
7. $x = -3$ $\quad y = 4$
8. $x = -3\frac{1}{2}$ $\quad y = 6$
9. $x = 5$ $\quad y = -2$
10. $x = -2$ $\quad y = -7$
11. $x = 2$ $\quad y = 5$
12. $x = 6$ $\quad y = 3$
13. $x = 4$ $\quad y = -2$
14. $x = -3$ $\quad y = 5$
15. $x = -2$ $\quad y = -3$
16. $x = 4$ $\quad y = -2$
17. $x = -1$ $\quad y = 3$
18. $x = -2$ $\quad y = -4$
19. $x = 3$ $\quad y = -2$
20. $x = -3$ $\quad y = -4$
21. $x = 2$ $\quad y = -6$
22. $x = -3$ $\quad y = 2$
23. $x = 1$ $\quad y = -4$
24. $x = -2$ $\quad y = -3$
25. $x = 5$ $\quad y = -2$

Exercise 12A

1. 2
2. 5
3. – 1
4. 3

5. 4
6. 0
7. 2
8. 3
9. 6
10. – 3

Exercise 12B

1. $x + y$
2. $x - y$
3. $2x + y$
4. $y - 3z$
5. $\frac{1}{2}(x - y)$ or $\frac{1}{2}x - \frac{1}{2}y$
6. $\frac{1}{2}y - \frac{1}{2}z$
7. $3x + y - z$
8. $x + 4y - 5z$
9. $\frac{1}{2}x - \frac{3}{2}z$
10. $\frac{1}{2}x + \frac{1}{4}z - \frac{1}{4}y$

Exercise 12C

1. 2
2. 4
3. 6
4. 3
5. 5
6. 2
7. 3
8. 5
9. 2
10. 4
11. 2
12. 6

Exercise 12D

1. 81
2. 256
3. 25
4. 343
5. 9
6. 8
7. 16
8. 36
9. 125
10. 64
11. 0.2 or $\frac{1}{5}$
12. 32
13. 216
14. $\frac{1}{9}$
15. 8

Exercise 12E

1. 2
2. 3
3. 4
4. 3
5. 5
6. 2
7. 2
8. 2 ½
9. 2
10. ⅔
11. 3

Exercise 12F

1. (i) p + q (ii) p – q (iii) 3q
(iv) 2p + 3q
2. (i) q – p (ii) 2p + q (iii) 4p
(iv) p + 2q
3. (i) p + q (ii) 2p + q (iii) 3p (iv) 2q
4. (i) p + q (ii) 5p (iii) 4p + q
(iv) q – p
5. (i) p + q (ii) q – p (iii) 2p + 2q (iv) 3q
6. (i) p + q (ii) q – p (iii) 6p
(iv) 2p + q
7. (i) p + q (ii) p – q (iii) p + 2q
(iv) 3p

Exercise 12G

1. (i) p + q (ii) 2p + q (iii) p + 2q
   (iv) 1 + p (v) 2 + q (vi) q – p
2. (i) p + q (ii) 2p + q (iii) p + 2q (iv) 3 + p (v) 2 + q
3. (i) q – p (ii) 2p + q (iii) p + 2q
   (iv) 1 + p (v) 2 + q (vi) 1 + p + q
4. (i) p + q (ii) q – p (iii) 2p + q
   (iv) 1 + p (v) 2 + q
5. (i) p + q (ii) q – p (iii) 1 + p
   (iv) 2p + q (v) p + 2q (vi) 1 + p + q
6. (i) p + q (ii) 1 + p (iii) 1 + q
   (iv) 2p + q (v) p – q
7. (i) p + q (ii) 1 + p (iii) 1 + q
   (iv) 2p + q (v) 1 – 2p
8. (i) 1 + p (ii) 2 + q (iii) q – p
   (iv) 2p + q (v) q – 2p (vi) 1 + p + q
9. (i) p + q (ii) 2 + p (iii) 1 + q (iv) q – p (v) 1 + p + q

Exercise 13A

1. 1.40
2. –0.827
3. 0.683
4. 0.631
5. –1.66

6. 1.51
7. 1.44
8. –2.11
9. 2.52
10. 5.13
11. –1.89
12. 0.648

Exercise 13B

1. –0.290
2. 0.658
3. –0.661
4. 4.21
5. 0.295
6. –15.4
7. 0.815
8. 0.127
9. –0.0155
10. 2.89

Exercise 13C

1. 2.83
2. –5.65
3. –1.10
4. –2.44
5. –0.287
6. 2.88
7. 2.49
8. 1.02
9. –0.154
10. –87.8

Exercise 14A

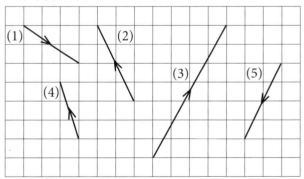

6. $\begin{pmatrix} 4 \\ 2 \end{pmatrix}$

7. $\begin{pmatrix} -2 \\ -3 \end{pmatrix}$

8. $\begin{pmatrix} 4 \\ 1 \end{pmatrix}$

9. $\begin{pmatrix} 0 \\ -3 \end{pmatrix}$

10. $\begin{pmatrix} -3 \\ 1 \end{pmatrix}$

11. $\begin{pmatrix} 3 \\ 0 \end{pmatrix}$

12. $\begin{pmatrix} -1 \\ 2 \end{pmatrix}$

Exercise 14B

1.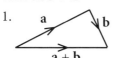

2.

3.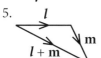

4.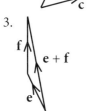

5.

6.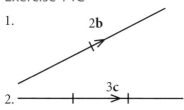

Exercise 14C

1.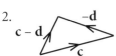
   2b

2. 3c

3. ½d

4.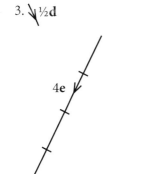
   4e

5.
   1½f

6.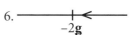
   −2g

7. −½h

8. −p

Exercise 14D

1. (i)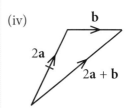
   a   b   a + b
   (ii) −b   a   a − b
   (iii) b   a   b − a

   (iv) 2a   b   2a + b

2. (i) c + d   d   c
   (ii) −c   d   d − c

   (iii)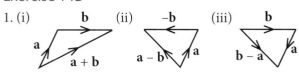
   c   −d   c − d
   (iv) −2c   d   d − 2c

3. (i) p   q   p + q
   (ii) q − p   q   −p

   (iii)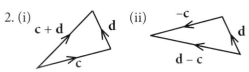
   p − q   −q   p

   (iv)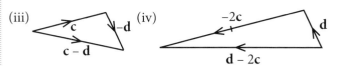
   2p   3q   2p + 3q

   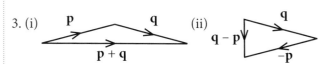

   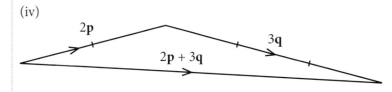

4. (i) 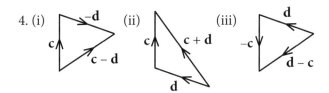 (ii) (iii)

(iv)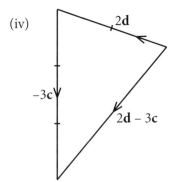

8. (i) 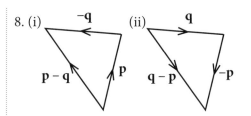 (ii)

(iii)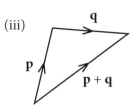

(iv)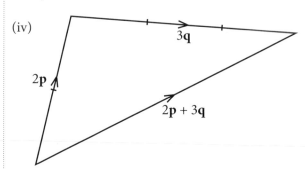

5. (i) 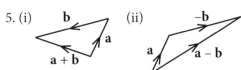 (ii)

(iii) (iv)

9. (i)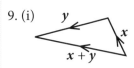

(ii) 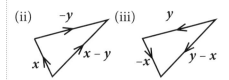 (iii)

6. (i) 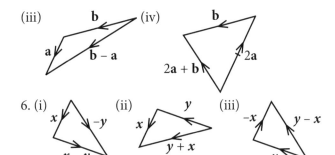 (ii) (iii)

(iv)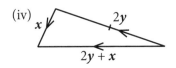

(iv)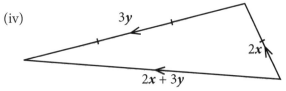

7. (i)

(ii) (iii)

(iv)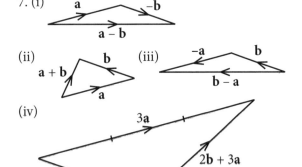

10. (i) 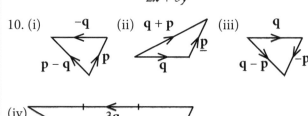 (ii) (iii)

(iv)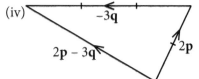

Exercise 14E

1. (a) (i) $-\boldsymbol{x} + \boldsymbol{y}$ (ii) $\frac{1}{2}\boldsymbol{x} + \frac{1}{2}\boldsymbol{y}$
   (b) (i) $1\frac{1}{2}\boldsymbol{y}$ (ii) $-\boldsymbol{x} + 1\frac{1}{2}\boldsymbol{y}$
2. (a) (i) $2\mathbf{a} + 3\mathbf{b}$ (ii) $-\mathbf{a} + 1\frac{1}{2}\mathbf{b}$
   (b) (i) $\frac{3}{4}\mathbf{b}$ (ii) $\mathbf{a} - \frac{3}{4}\mathbf{b}$
3. (a) (i) $\mathbf{a} + 3\mathbf{b}$ (ii) $-3\mathbf{b} + \mathbf{a}$
   (b) (i) $3\mathbf{a}$ (ii) $3\mathbf{a} + 3\mathbf{b}$
4. (a) (i) $-\mathbf{a}+\mathbf{b}$ (ii) $\mathbf{b} - \frac{1}{2}\mathbf{a}$
   (b) (i) $-\frac{2}{3}\mathbf{a} + \frac{2}{3}\mathbf{b}$ (ii) $-\frac{1}{6}\mathbf{a} - \frac{1}{3}\mathbf{b}$
5. (i) $4\mathbf{p}$ (ii) $2\mathbf{p} - 3\mathbf{q}$
   (iii) $6\mathbf{q} - 2\mathbf{p}$ (iv) $4\mathbf{p} - 6\mathbf{q}$
6. (a) (i) $-4\boldsymbol{x} + 3\boldsymbol{y}$ (ii) $2\boldsymbol{x} + 1\frac{1}{2}\boldsymbol{y}$
   (b) (i) $9\boldsymbol{y}$ (ii) $2\boldsymbol{x} - 7\frac{1}{2}\boldsymbol{y}$
7. (a) (i) $\mathbf{b} - \mathbf{a}$ (ii) $\frac{1}{2}\mathbf{a} + \frac{1}{2}\mathbf{b}$
   (b) (i) $\frac{2}{5}\mathbf{a}$ (ii) $\frac{1}{2}\mathbf{b} - 1\frac{1}{10}\mathbf{a}$
8. (a) (i) $4\boldsymbol{y}$ (ii) $\boldsymbol{y} - 2\boldsymbol{x}$ (iii) $3\frac{1}{2}\boldsymbol{y} + \boldsymbol{x}$
   (b) (i) $12\boldsymbol{y}$ (ii) $8\frac{1}{2}\boldsymbol{y} - \boldsymbol{x}$
9. (i) $2\mathbf{a}$ (ii) $2\mathbf{a} - 2\mathbf{b}$
   (iii) $2\mathbf{a} - 4\mathbf{b}$ (iv) $4\mathbf{b}$
10. (a) (i) $2\mathbf{b}$ (ii) $\mathbf{a} - \mathbf{b}$ (iii) $\mathbf{a} + \mathbf{b}$
    (b) (i) $\mathbf{b}$ (ii) $-\mathbf{a}$

Exercise 14F

1. (a) (i) $2\mathbf{p} + 3\mathbf{q}$ (ii) $-2\mathbf{p} + 3\mathbf{q}$
   (b) $\overrightarrow{AM} = \frac{1}{2}\overrightarrow{AC}$ Both vectors share point A and are parallel, therefore they must be colinear.
2. (a) $-\boldsymbol{y} + \boldsymbol{x}$
   (b) $\overrightarrow{EF} = 3\boldsymbol{x} - 3\boldsymbol{y} = 3(\boldsymbol{x} - \boldsymbol{y}) = 3\overrightarrow{DC}$
3. (a) (i) $6\mathbf{q}$ (ii) $2\mathbf{p} - 2\mathbf{q}$ (iii) $-4\mathbf{q} + 4\mathbf{p}$
   (b) MN = 2DC and they are parallel
4. (a) (i) $3\mathbf{a}$ (ii) $-3\mathbf{a} + 4\mathbf{b}$ (iii) $-\frac{4}{3}\mathbf{b} + \mathbf{a}$
   (b) QR = 3TU and they are parallel
5. (a) (i) $-3\mathbf{t} + 2\mathbf{v}$ (ii) $-3\mathbf{t}$ (iii) $-6\mathbf{t} + 2\mathbf{v}$ (iv) $-6\mathbf{t} + 2\mathbf{v}$
   (b) DB = BG and they are collinear

Exercise 15A

1. (i) 3 m/s
   (ii) 0 m/s
   (iii) 3 m/s in the opposite direction
2. (i) 6 m/s
   (ii) 0 m/s
   (iii) 4 m/s in the opposite direction
3. (i) 1.6 m/s
   (ii) 0 m/s
   (iii) 2 m/s in the opposite direction
4. (i) 11 m/s
   (ii) 0 m/s
   (iii) 6.25 m/s in the opposite direction
5. (i) 6.8 m/s
   (ii) 2 m/s in the opposite direction
   (iii) 7 m/s in the opposite direction
6. (i) 2.4 m/s
   (ii) $2\frac{1}{3}$ m/s in the opposite direction
   (iii) 3.5 m/s in the opposite direction

Exercise 15B

1. (i) 1.5 m/s^2, 0 m/s^2, −4 m/s^2 (ii) 138 m
2. (i) 1.6 m/s^2, 0 m/s^2, −2 m/s^2 (ii) 92 m
3. (i) 2 m/s^2, 0 m/s^2, −2.5 m/s^2 (ii) 111 m
4. (i) 1.6 m/s^2, 0 m/s^2, −3.8 m/s^2 (ii) 423.5 m
5. (i) 2.4 m/s^2, −2 m/s^2 (ii) 264 m
6. (i) 0.8 m/s^2, −0.9 m/s^2 (ii) 80 m
7. (i) 1.6 m/s^2, 0 m/s^2, −2 m/s^2 (ii) 140 m
8. (i) 2.5 m/s^2, 0 m/s^2, −9 m/s^2 (ii) 371.25 m
9. (i) $1\frac{2}{3}$ m/s^2, 0 m/s^2, −5 m/s^2 (ii) 540 m
10. (i) $\frac{2}{3}$ m/s^2, 0 m/s^2, −2.6 m/s^2 (ii) 348.5 m
11. (a)

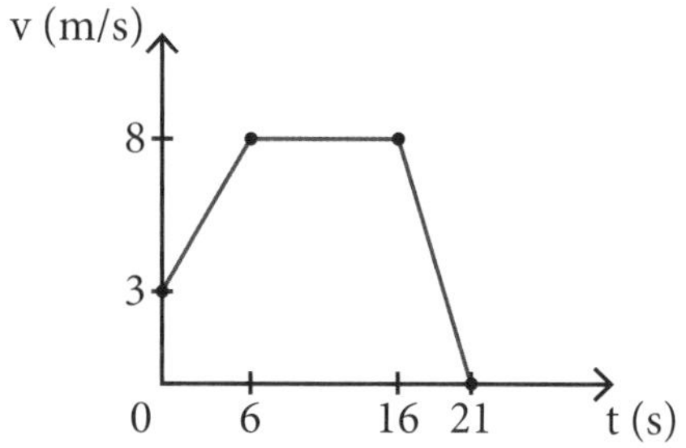

(b)(i) $\frac{5}{6}$ m/s^2 (ii) 1.6 m/s^2 (iii) 133 m

12. (a)

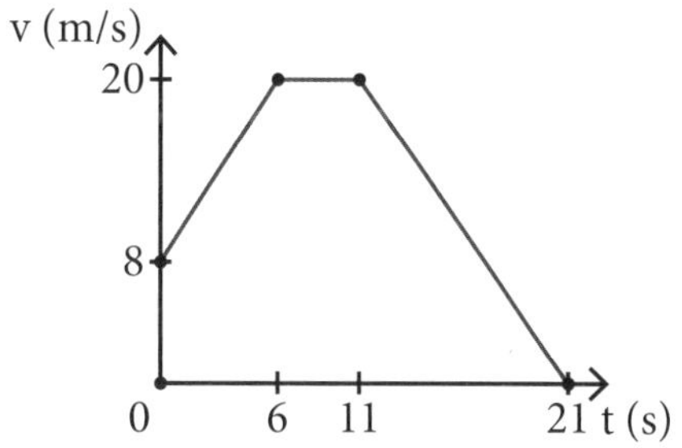

(b)(i) 2 m/s^2 (ii) 2 m/s^2 (iii) 284 m

13. (a)

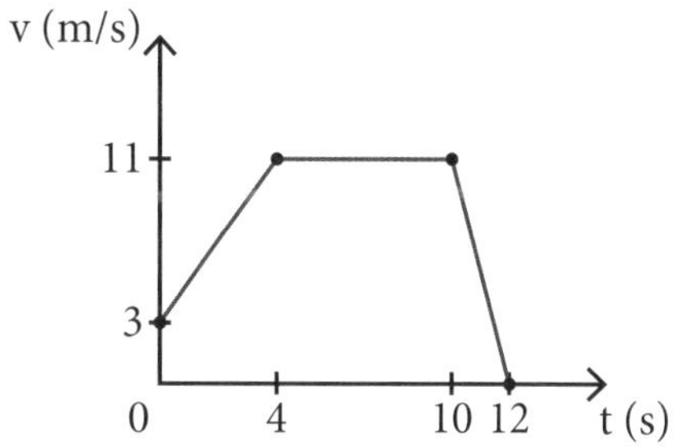

(b)(i) 2 m/s^2 (ii) 5.5 m/s^2 (iii) 105 m

14. (a)

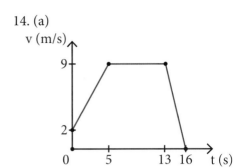

(b)(i) 1.4 m/s² (ii) 3 m/s² (iii) 113 m

15. (a)

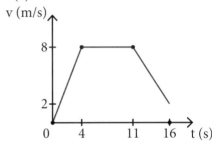

(b)(i) 2 m/s² (ii) 1.2 m/s² (iii) 97 m

16. (a)

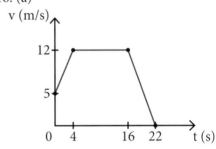

(b)(i) 1.75 m/s² (ii) 2 m/s² (iii) 214 m

17. (a)

(b)(i) 4 m/s² (ii) 1 ⁷⁄₉ m/s² (iii) 318 m

Exercise 15C

1. (a)

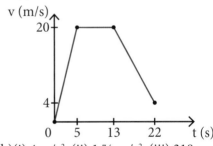

(b) (i) 8 m/s (ii) 345.9 m (iii) 0.4 m/s² (iv) 56 m

2. (a)

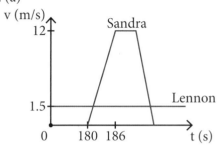

(b) (i) 209 s (ii) 313.7 m (iii) 45 m

Exercise 16A

1. (i) 1.5 m/s²
   (ii) 24m
2. (i) ⅔ m/s²
   (ii) 24m
3. (i) 14 m/s
   (ii) 44 m
4. (i) 15 m/s
   (ii) 0.625 m/s²
5. (i) 4 s
   (ii) 2.25 m/s²
6. (i) 80 m
   (ii) 10 s
7. (i) 2 m/s
   (ii) ⅙ m/s²
8. (i) 22 m/s
   (ii) 78 m
9. (i) 3 m/s
   (ii) 8 s
10. (i) 6 m/s²
    (ii) 32 m/s
11. (i) 33.5 m/s
    (ii) 9 s
12. (i) 7 m/s
    (ii) 17 m
13. (i) 3 m/s²
    (ii) 7 s
14. (i) 3.55 m/s²
    (ii) 139.425 m

Exercise 16B

1. (i) 6 m/s
   (ii) 18m
2. (i) 17 m/s
   (ii) 4 s
3. (i) 3 m/s²
   (ii) 4 s
4. (i) 1.5 s
   (ii) 12.75 m

5. (i) 1.5 m/s^2
   (ii) 3 m
6. (i) 8 m/s
   (ii) 3 s
7. (i) 8 m/s
   (ii) 7.5 m
8. (i) 24 m/s
   (ii) 12 m/s^2

Exercise 16C

1. (i) 25 m/s
   (ii) 31.25 m
2. (i) 0.098 m
   (ii) 0.14 s
3. (i) 6.07 m/s
   (ii) 0.607 s
4. (i) 0.2645 m
   (ii) 0.46 s
5. (i) 26.3 m
   (ii) 131.5 m
   (iii) 51.3 m/s
6. (i) 0.722 m
   (ii) 0.169 s and 0.591 s
7. (i) 0.736 s
   (ii) 35.5 m
8. (i) 0.44 m/s going down
   (ii) 0.156 s
   (iii) 0.122 m
   (iv) 0.0647 s and 0.247 s

Exercise 16D

1. (i) 0.264 s
   (ii) 1.05 m
   (iii) 0.605 m
2. (i) 0.87 s
   (ii) 3.78 m
   (iii) 8.7 m/s and 9 m/s
3. (i) 4.05 m
   (ii) 1.4 s
   (iii) 9 m/s
   (iv) 0.168 s and 0.832 s
4. (i) 8.912 m
   (ii) 1.66 s
   (iii) 13.4 m/s
   (iv) 0.07 s to 0.57 s
5. (i) 0.523 s
   (ii) 0.932 m
   (iii) 0.422 s
6. (i) 0.625 s
   (ii) 3.05 m
   (iii) 1.8 m

Exercise 17A

1. 20 N
2. 6 kg
3. 5 m/s^2
4. 24 N
5. 8 kg
6. 1.5 m/s^2
7. 22.4 N
8. 30 kg

Exercise 17B

1. (i) 6 N
   (ii) 51 m
2. (i) 27 N
   (ii) 2 s
3. (i) 27 N
   (ii) 7 m/s
4. (i) 56 N
   (ii) 37 m/s
5. (i) 24 N
   (ii) 6 s
6. (i) 22.4 N
   (ii) 37.5 m
7. (i) 24 N
   (ii) 39 m/s
8. (i) 37.5 N
   (ii) 8 m/s

Exercise 17C

1. (i) 14 N
   (ii) 4 m/s
2. (i) 16 N
   (ii) 1.5 s
3. (i) 36 N
   (ii) 18 m
4. (i) 21 N
   (ii) 7 m/s
5. (i) 10 N
   (ii) 4 s
6. (i) 10.5 N
   (ii) 2 s
7. (i) 12 N
   (ii) 13.5 m
8. (i) 20 N
   (ii) 5 m/s
9. (i) 9 N
   (ii) 4 s
10. (i) 12.6 N
    (ii) 17.5 m

Exercise 18A

1. Horizontal component = 22.6 N
   Vertical component = 8.21 N
2. Horizontal component = 0.923 N
   Vertical component = 2.54 N
3. Horizontal component = 11.1 N
   Vertical component = 11.5 N
4. Horizontal component = 11.8 N
   Vertical component = 2.08 N
5. Horizontal component = 6.14 N
   Vertical component = 12.6 N
6. Horizontal component = 20.1 N
   Vertical component = 28.7 N
7. Horizontal component = 43.7 N
   Vertical component = 31.7 N
8. Horizontal component = 21.1 N
   Vertical component = 39.7 N
9. Horizontal component = 2.82 N
   Vertical component = 1.9 N
10. Horizontal component = 41.8 N
    Vertical component = 32.6 N

Exercise 18B

1. 21.6 N at 87.6° with the horizontal
2. 3.23 N at 47.7° with the horizontal
3. 29.4 N at 42.0° with the horizontal
4. 45.8 N at 20.8° with the horizontal
5. 2.87 N at 76.1° with the horizontal
6. 240 N at 24.6° with the horizontal
7. 3.37 N at 64.1° with the horizontal
8. 50.1 N at 3.58° with the horizontal
9. 19.6 N at 12.6° with the horizontal
10. 4.06 N at 41.5° with the horizontal

Exercise 18C

1. 13 N at 32.5° with the horizontal
2. 12.1 N at 65.6° with the horizontal
3. 3.61 N at 56.3° with the horizontal
4. 55.1 N at 4.16° with the horizontal
5. 9.22 N at 77.5° with the horizontal
6. 4.27 N at 32.6° with the horizontal
7. 26.4 N at 65.4° with the horizontal
8. 28.4 N at 27.6° with the horizontal
9. 12.5 N at 28.6° with the horizontal
10. 17.7 N at 42.7° with the horizontal

Exercise 18D

1. 19.9 N at 21.0° with the inclined plane
2. 7.53 N at 7.7° with the inclined plane
3. 13.8 N at 24.4° with the inclined plane
4. 196 N at 6.56° with the inclined plane
5. 2.56 N at 22.8° with the inclined plane
6. 99.5 N at 11.4° with the inclined plane

Exercise 18E

1. A = 1.38 N   B = 7.92 N
2. C = 38.1 N   D = 28.8 N
3. V = 52.6 N   W = 16.2 N
4. F = 18.7 N   G = 17.6 N
5. H = 13.9 N   I = 15.8 N
6. L = 54.3 N   M = 66.3 N
7. P = 49.7 N   Q = 12.9 N
8. R = 3.07 N   S = 4.92 N
9. U = 56.4 N   V = 77.7 N
10. X = 50.4 N   Y = 19.4 N

Exercise 18F

1. R = 16.7 N   Q = 19.9 N
2. R = 5.91 N   Q = 4.45 N
3. R = 27.5 N   Q = 39.3 N
4. R = 1.33 N   Q = 2.51 N
5. R = 283 N   Q = 138 N
6. R = 10.2 N   Q = 33.5 N

Exercise 18G

1. R = 26.6 N   Q = 25.6 N
2. R = 2.78 N   Q = 3.41 N
3. R = 233.8 N Q = 294.8 N
4. R = 63.8 N   Q = 61.5 N
5. R = 7.63 N   Q = 5.91 N
6. R = 46.3 N   Q = 53.6 N

Exercise 19A

1. 6.4 N at 51.3° with the horizontal
2. 7.28 N at 15.9° with the horizontal
3. 6.71 N at 63.4° with the horizontal
4. 17 N at 61.9° with the horizontal
5. 3.16 N at 71.6° with the horizontal
6. 5.39 N at 21.8° with the horizontal
7. 13 N at 22.6° with the horizontal
8. 4.47 N at 26.6° with the horizontal
9. 7.81 N at 50.2° with the horizontal
10. 7.62 N at 23.2° with the horizontal

Exercise 19B

1. 10 m/s
2. 8.06 m/s
3. 8.25 m/s
4. 2.24 m
5. 26 m
6. 11.4 m

Exercise 19C

1. $(6\mathbf{i} + 2\mathbf{j})$ N
2. $(-5\mathbf{i} + 2\mathbf{j})$ N
3. $(-5\mathbf{i} - \mathbf{j})$ N
4. $(\mathbf{j})$ N
5. $(3\mathbf{i})$ N
6. $(4\mathbf{i} - \mathbf{j})$ N
7. $(-2\mathbf{i} - 4\mathbf{j})$ N
8. $(-3\mathbf{i} - 2\mathbf{j})$ N
9. $(6\mathbf{i} + 3\mathbf{j})$ N
10. $(\mathbf{i} + 12\mathbf{j})$ N

Exercise 19D

1. $(\mathbf{i} - 2\mathbf{j})$ N
2. $(-3\mathbf{i} + 4\mathbf{j})$ N
3. $(\mathbf{i} + 5\mathbf{j})$ N
4. $(-4\mathbf{i} - 2\mathbf{j})$ N
5. $(3\mathbf{i} + 8\mathbf{j})$ N
6. $(-2\mathbf{i} + 3\mathbf{j})$ N
7. $(5\mathbf{i} - \mathbf{j})$ N
8. $(-2\mathbf{j})$ N
9. $(4\mathbf{i} + 3\mathbf{j})$ N
10. $(-3\mathbf{i} - \mathbf{j})$ N

Exercise 19E

1. $(\mathbf{i} - 3\mathbf{j})$ m/s^2
2. $(-\mathbf{i} + 4\mathbf{j})$ N
3. $(-\mathbf{i} - 2\mathbf{j})$ m/s^2
4. $(-2\mathbf{i} - 6\mathbf{j})$ N
5. $(-2\mathbf{i} + 3\mathbf{j})$ m/s^2
6. $(-3\mathbf{i} + 8\mathbf{j})$ N
7. $(-\mathbf{i} + 3\mathbf{j})$ m/s^2
8. $(-3\mathbf{i} - 5\mathbf{j})$ N
9. $(3\mathbf{i} + 2\mathbf{j})$ m/s^2
10. $(-2\mathbf{i} + 5\mathbf{j})$ N

Exercise 19F

1. (i) $(4\mathbf{i} - 12\mathbf{j})$ N (ii) $(6\mathbf{i} - 11\mathbf{j})$ m/s (iii) $(13.5\mathbf{i} - 19.5\mathbf{j})$ m
2. (i) $(2\mathbf{i} + 2\mathbf{j})$ m/s^2 (ii) $(3\mathbf{i} + 6\mathbf{j})$ m/s (iii) $(2\mathbf{i} + 8\mathbf{j})$ m
3. (i) $(2\mathbf{i} + 6\mathbf{j})$ N (ii) $(-2\mathbf{i} - 4\mathbf{j})$ m/s (iii) $(6\mathbf{i} + 30\mathbf{j})$ m
4. (i) $-7.5\mathbf{j}$ m/s^2 (ii) $(4\mathbf{i} - 33\mathbf{j})$ m/s (iii) $(16\mathbf{i} - 72\mathbf{j})$ m
5. (i) $(-12\mathbf{i} - 30\mathbf{j})$ N (ii) $(3\mathbf{i} + 4\mathbf{j})$ m/s (iii) $-10.5\mathbf{j}$ m
6. (i) $(2.5\mathbf{i} + 12.5\mathbf{j})$ m/s^2 (ii) $(10.5\mathbf{i} + 57.5\mathbf{j})$ m/s
   (iii) $(21.25\mathbf{i} + 131.25\mathbf{j})$ m
7. (i) $(-14\mathbf{i} + 7\mathbf{j})$ N (ii) $(4\mathbf{i} - 3\mathbf{j})$ m/s (iii) $(-4\mathbf{i} + \mathbf{j})$ m/s
8. (i) $(4\mathbf{i} + 2\mathbf{j})$ m/s^2 (ii) $(7\mathbf{i} + 7\mathbf{j})$ m/s (iii) $(6\mathbf{i} + 10\mathbf{j})$ m
9. (i) $(-8\mathbf{i} + 16\mathbf{j})$ N (ii) $(5\mathbf{i} - 2\mathbf{j})$ m/s (iii) $(\mathbf{i} + 6\mathbf{j})$ m/s
10. (i) $(-30\mathbf{i} + 18\mathbf{j})$ m/s^2 (ii) $(172\mathbf{i} - 96\mathbf{j})$ m/s
    (iii) $(492\mathbf{i} - 252\mathbf{j})$ m

Exercise 19G

1. $(10\mathbf{i} - 24\mathbf{j})$ m/s
2. $(-18\mathbf{i} + 24\mathbf{j})$ m/s
3. $(45\mathbf{i} + 24\mathbf{j})$ m/s
4. $(-21\mathbf{i} - 72\mathbf{j})$ m/s
5. $(36\mathbf{i} + 27\mathbf{j})$ m/s
6. $(-36\mathbf{i} + 15\mathbf{j})$ m/s

Exercise 20A

1. 0.625
2. 0.643
3. 182.4 N
4. 34.56 N
5. 3.29 kg
6. 1.67 kg
7. 0.521
8. 67.2 N
9. 3.72 kg
10. 190.4 N

Exercise 20B

1. 0.364
2. 0.9
3. (i) 33.2 N (ii) 15.6 N
4. (i) 58.7 N (ii) 26.1 N
5. (i) 23.7 N (ii) 5.3 N
6. (i) 25.5 N (ii) 6.66 N
7. (i) 99.2 N (ii) 5.71 N
8. (i) 26.9 N (ii) 5.58 N
9. (i) 20.4 N (ii) 0.389 N
10. (i) 78.5 N (ii) 8.70 N

Exercise 20C

1. (i) 9.65 m/s^2 (ii) 0.259 s
2. (i) 0.367 (ii) 112m
3. (i) 7.2 m/s^2 (ii) 1.22 s
4. (i) 0.729 (ii) 12 m/s
5. (i) 1.53 m/s^2 (ii) 19.2 m
6. (i) 0.709 (ii) 10 s
7. (i) 0.508 m/s^2 (ii) 2.89 m/s
8. (i) 0.772 (ii) 3 m/s
9. (i) 1.63 m/s^2 (ii) 6.53 m
10. (i) 1.5 m/s^2 (ii) 18.75 m/s
11. (i) 0.498 (ii) 77 m/s

Exercise 20D

1. 0.780
2. 0.162
3. 71.22 N
4. 56.4 N

5. 0.31
6. 0.155
7. 69.3 N
8. 0.142
9. 59.1 N
10. 0.520

Exercise 21A

1. (i) 1.18 m/s^2 (ii) 462 kg
2. (i) 1.45 m/s^2 (ii) 1079 N
3. (i) 837 N (ii) 1422 N
4. (i) 1423.75 N (ii) 1056.25 N
5. (i) 3500 N (ii) 254 kg
6. (i) 1440 kg (ii) 277 kg
7. (i) 1.34 m/s^2 (ii) 347.2 N
8. (i) 1505 kg (ii) 785 N
9. 25.3 N
10. 0.399
11. 9.38 kg
12. 3.32 m/s^2

Exercise 21B

1. (i) 1.2 m/s^2 (ii) 1094 N
2. (i) 1.32 m/s^2 (ii) 1084 N
3. (i) 1124.68 N (ii) 1.45 N per kg
4. (i) 861.12 N (ii) 1.66 N per kg
5. (i) 3370.72 N (ii) 442 kg
6. (i) 1165 kg (ii) 429 kg
7. (i) 1.3 m/s^2 (ii) 0.77 N per kg
8. (i) 1275 kg (ii) 748.84 N

Exercise 21C

1. (a) 2253 N (b) 378 N (c)(i)2545.5 N (c)(ii) 445.5 N
   (d) 0.84 m/s^2
2. (a) 2403 N (b) 513 N (c)(i) 2970 N (c) (ii)675 N
   (d) 0.95 m/s^2
3. (a) 4620 N (b) 2700 N (c) (i)5376 N (c) (ii)3120 N
   (d) 1.8 m/s^2
4. (a) 4670 N (b) 2800 N (c) 7.2 m (d) 2 m/s
   (e) 1.25 m
5. (i) 1.25 m/s^2 (ii) 1643.5 N (iii) 912.6 N (iv) 730.9 N
   (v) 1380.9 N (vi) 14.2 s
6. (i) 1.6 m/s^2 (ii) 1471.52 N (iii) 1176 N (iv) 4089.12 N
   (v) 7.06 s (vi) 21.18 m

Exercise 21D

1. (i) 3.85 m/s^2 (ii) 5.54 N (iii) 11.68 N
2. (i) 2 m/s^2 (ii) 9.6 N (iii) 19.2 N
3. (i) 2 m/s^2 (ii) 2.4 N (iii) 4.8 N
4. (i) 1 ⅔ m/s^2 (ii) 5.83 N (iii) 11.7 N

5. (i) 2 m/s^2 (ii) 12 N (iii) 24 N
6. (i) 3.75 m/s^2 (ii) 6.875 N (iii) 13.75 N
7. (i) 2.5 m/s^2 (ii) 3.75 N (iii) 7.5 N
8. (i) 4 m/s^2 (ii) 42 N (iii) 84 N

Exercise 21E

1. (i) 2 m/s^2 (ii) 4.8 N (iii) 9.6 N (iv) 1.1 s (v) 2.2 m/s
   (vi) 0.242 m (vii) 0.44 s
2. (i) 3 m/s^2 (ii) 18.2 N (iii) 36.4 N (iv) 2.16 m (v) 3.6 m/s
   (vi) 0.648 m (vii) 0.72 s
3. (i) 4 m/s^2 (ii) 4.2 N (iii) 8.4 N (iv) 4.5 m (v) 6 m/s
   (vi) 1.8 m (vii) 1.2 s
4. (i) 4.4 m/s^2 (ii) 20.16 N (iii) 40.32 N (iv) 0.58 s
   (v) 2.57 m/s (vi) 0.33 m (vii) 0.514 s
5. (i) 3 m/s^2 (ii) 9.1 N (iii) 18.2 N (iv) 0.96 m (v) 2.4 m/s
   (vi) 0.288 m (vii) 0.48 s
6. (i) 4.8 m/s^2 (ii) 19.24 N (iii) 38.48 N (iv) 0.384 m
   (v) 1.92 m/s (vi) 0.384 s
7. (i) 5 m/s^2 (ii) 7.5 N (iii) 15 N (iv) 1.2 s (v) 3.6 m
   (vi) 1.2 s
8. (i) 6 m/s^2 (ii) 9.6 N (iii) 19.2 N (iv) 0.5 s (v) 0.75 m
   (vi) 0.6 s

Exercise 21F

1. (i) 5.33 m/s^2 (ii) 25.7 N (iii) 52.8 N
2. (i) 7.14 m/s^2 (ii) 14.3 N (iii) 20.2 N
3. (i) 7.27 m/s^2 (ii) 21.8 N (iii) 30.8 N
4. (i) 6.36 m/s^2 (ii) 2.54 N (iii) 3.59 N
5. (i) 5.71 m/s^2 (ii) 13.7 N (iii) 19.4 N
6. (i) 5.79 m/s^2 (ii) 92.6 N (iii) 131 N
7. (i) 5.71 m/s^2 (ii) 25.7 N (iii) 36.4 N
8. (i) 6 m/s^2 (ii) 4.8 N (iii) 6.79 N
9. (i) 5.625 m/s^2 (ii) 39.375 N (iii) 55.7 N
10. (i) 5.56 m/s^2 (ii) 267 N (iii) 377 N

Exercise 21G

1. (i) 4.86 m/s^2 (ii) 35.7 N (iii) 2.33 m/s (iv) 0.291 s
   (v) 0.339 m
2. (i) 5 m/s^2 (ii) 3.5 N (iii) 2.53 m/s (iv) 0.675 s (v) 0.853 m
3. (i) 3.6 m/s^2 (ii) 153.6 N (iii) 1.51 m/s (iv) 0.252 s
   (v) 0.19 m
4. (i) 4.13 m/s^2 (ii) 10.6 N (iii) 3.15 m/s (iv) 0.7875 s
   (v) 1.24 m
5. (i) 4.43 m/s^2 (ii) 334 N (iii) 1.06 m/s (iv) 0.353 s
   (v) 0.187 m
6. (i) 4 m/s^2 (ii) 126 N (iii) 1.90 m/s (iv) 0.633 s
   (v) 0.602 m
7. (i) 3.32 m/s^2 (ii) 4.81 N (iii) 2.52 m/s (iv) 0.6 s
   (v) 0.756 m
8. (i) 4.17 m/s^2 (ii) 69.9 N (iii) 2.66 m/s (iv) 0.739 s
   (v) 0.983 m

9. (i) 3.41 m/s² (ii) 276.9 N (iii) 2.59 m/s (iv) 0.604 s
   (v) 0.783 m
10. (i) 2.21 m/s² (ii) 22.6 N (iii) 2.26 m/s (iv) 0.314 s
    (v) 0.355 m

Exercise 22A

1. 32 N and 48 N
2. 22.7 N and 27.3 N
3. 33.75 N and 26.25 N
4. 45 N and 75 N
5. 20.16 N and 35.84 N
6. 10.3 N and 15.7 N
7. 3.48 N and 4.52 N
8. 15.2 N and 20.8 N

Exercise 22B

1. 21.4 N and 28.6 N
2. 23.5 N and 56.5 N
3. 15 N and 9 N
4. 22.4 N and 33.6 N
5. 2.57 N and 6.43 N
6. 9.85 N and 6.15 N
7. 5.2 N and 24.8 N
8. 127.1 N and 112.9 N
9. 0.94 N and 3.06 N
10. 23.5 N and 12.5 N

Exercise 22C

1. 6.13 m
2. 7.83 m
3. 4.71 m
4. 8.93 m
5. 4.59 m
6. 2.37 m

Exercise 22D

1. 16.8 kg
2. 0.5 m
3. 12.65 kg
4. 5.31 m
5. 1.91 kg
6. 6.27 m
7. (a) 72 N and 48 N (b) 6 kg

Exercise 23A

6. Discrete
7. Continuous
8. Continuous
9. Discrete
10. Continuous

15. (a) Discrete (b) Continuous (c) Continuous
    (d) Discrete
16. (a) Discrete (b) Continuous (c) Continuous
    (d) Continuous (e) Discrete
17. (i) Continuous (ii) Discrete (iii) Continuous
18. None left to sell
19. Sample 1 (i) Biased as they are studying Further Maths
    (ii) Choose 10 from every subject*
Sample 2 (i) Biased as only boys chosen (ii) Choose 30
    boys and 30 girls, one from every class*
Sample 3 (i) Biased as only senior pupils are chosen
    (ii) Choose 6 pupils, one from every class in school*
    *or a similar answer that removes the bias

Exercise 23B

1 (a) 19 (b) 18.5 (c) 9.5
2 (a) 1 (b) 19 (c) 41.5
3 (a) 44 (b) 31.5 (c) 22.5
4 (a) 76 (b) 40 (c) 70
5 (a) 4 (b) 9.5 (c) 5
6 (a) 13 (b) 19 (c) 34

Exercise 23C

1. Frequency Density: 4.2, 9.6, 1.5, 5.4

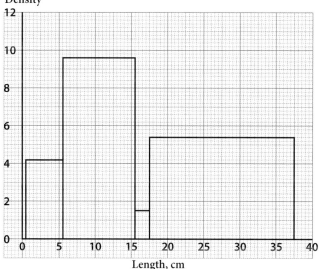

2. Frequency Density: 142, 89, 131, 48, 196

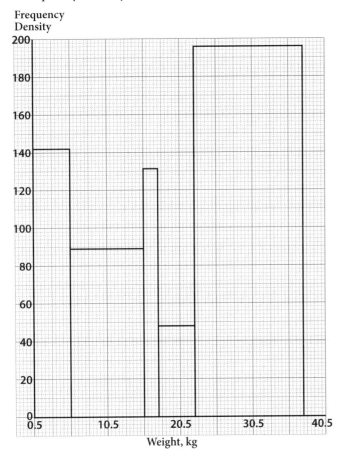

Weight, kg

3. Frequency Density: 34, 64, 92, 32

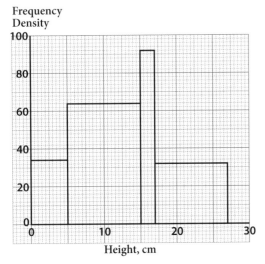

Height, cm

4. Frequency Density: 8.4, 3.6, 4.5, 8.4, 2.8

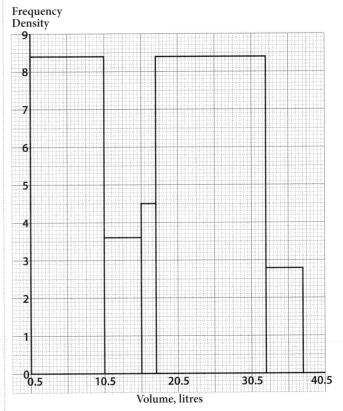

Volume, litres

5. Frequency Density: 12, 13, 8.6, 9.2

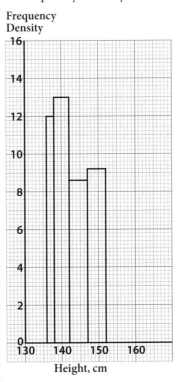

Height, cm

6. Frequency Density: 3.4, 2.2, 4.6, 2

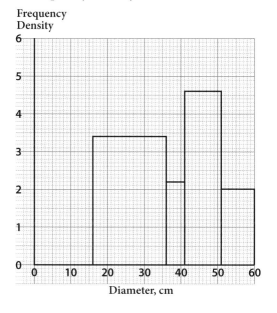

Frequency Density

Diameter, cm

7. Frequency Density: 84, 46, 72, 35

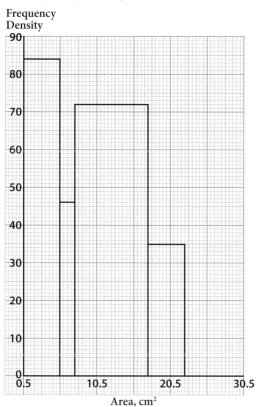

Frequency Density

Area, cm²

8. Frequency Density: 6.4, 3.8, 8.5, 2.7

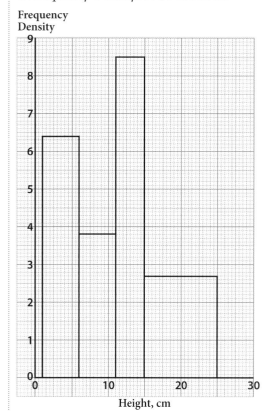

Frequency Density

Height, cm

9. Frequency Density: 26, 8, 14

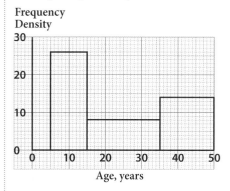

Frequency Density

Age, years

10. Frequency Density: 1.4, 2.5, 0.8

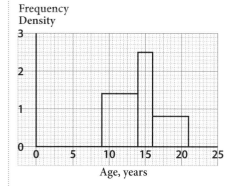

Frequency Density

Age, years

Exercise 23D

1. (a)

Length, cm	Frequency
1 – 5	18
6 – 8	18
9 – 18	82
19 – 23	22

(b) 9, 18 (c) 8.5, 18.5

2. (a)

Volume V, litres	Frequency
0 < V ≤ 10	230
10 < V ≤ 15	90
15 < V ≤ 35	220
35 < V ≤ 37	44
37 < V ≤ 42	215

(b) 0, 10 (c) 15, 35

3. (a)

Area A, m²	Frequency
0 < A ≤ 8	344
8 < A ≤ 13	105
13 < A ≤ 17	48
17 < A ≤ 27	320
27 < A ≤ 30	54

(b) 0, 8 (c) 8, 13

4. (a)

Width, cm	Frequency
1 – 10	34
11 – 15	11
16 – 17	3
18 – 37	32
38 – 42	14

(b) 1, 10 (c) 15.5, 17.5

5. (a)

Mass, kg	Frequency
1 – 4	14
5 – 19	96
20 – 29	74
30 – 34	28
35 – 36	17

(b) 5, 19 (c) 19.5, 29.5

6. (a)

Length l, cm	Frequency
4 < l ≤ 9	70
9 < l ≤ 11	14
11 < l ≤ 21	230
21 < l ≤ 26	240
26 < l ≤ 32	18

(b) 21, 26 (c) 11, 21

Exercise 23E

1. (a)

Mass, kg	Frequency
5 – 11	119
12 – 16	120
17 – 20	128
21 – 30	160
31 – 36	144

(b)

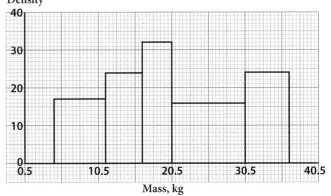

2. (a)

Circumference C, cm	Frequency
10 < C ≤ 14	14
14 < C ≤ 19	42
19 < C ≤ 29	26
29 < C ≤ 31	9
31 < C ≤ 36	9

(b)

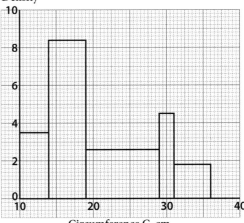

3. (a)

Length, m	Frequency
3 – 9	42
10 – 14	18
15 – 24	42
25 – 28	30
29 – 33	12

(b)

Frequency Density

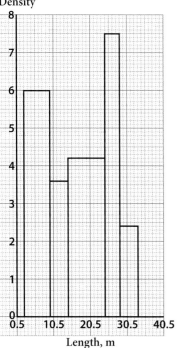

Length, m

4. (a)

Perimeter, cm	Frequency
12 – 16	29
17 – 20	14
21 – 22	13
23 – 32	23
33 – 37	18

(b)

Frequency Density

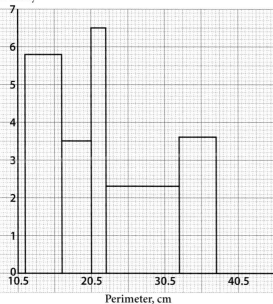

Perimeter, cm

5. (a)

Volume V, cm³	Frequency
$5 < V \leq 10$	42
$10 < V \leq 20$	37
$20 < V \leq 24$	26
$24 < V \leq 29$	21
$29 < V \leq 33$	10

(b)

Frequency Density

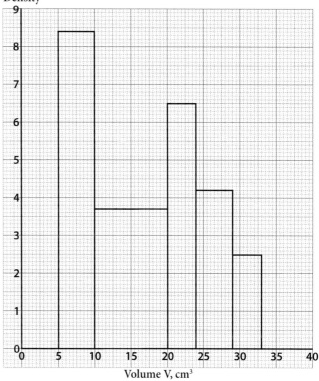

Volume V, cm³

Exercise 24A

1. Mean = 13.5 and Standard Deviation = 1.8
2. Mean = 6.72 and Standard Deviation = 1.62
3. Mean = 126.8 and Standard Deviation = 8.66
4. Mean = £5.67 and Standard Deviation = £2.14
5. Mean = 11.375 and Standard Deviation = 2.23
6. Mean = 8.25 and Standard Deviation = 2.49
7. Mean = 24.75 and Standard Deviation = 3.38
8. Mean = 20.3 and Standard Deviation = 3.74

Exercise 24B

1. Mean = 2 and Standard Deviation = 0.837
2. Mean = 4.47 and Standard Deviation = 1.77
3. Mean = 3.72 and Standard Deviation = 2.72
4. Mean = 2.02 and Standard Deviation = 0.927
5. Mean = 2.625 and Standard Deviation = 1.84
6. Mean = 5.64 and Standard Deviation = 2.41
7. Mean = 26.1875 and Standard Deviation = 2.02

8. Mean = 8.35 and Standard Deviation = 2.31
9. Mean = 6.84 and Standard Deviation = 1.39
10. Mean = 14.25 and Standard Deviation = 1.87

Exercise 24C

1. Mean = 9.9 and Standard Deviation = 5.39
2. Mean = 12.28 and Standard Deviation = 6.08
3. Mean = 6.1 and Standard Deviation = 2.05
4. Mean = 31.1 and Standard Deviation = 12.83
5. Mean = 12.8 and Standard Deviation = 6.8
6. Mean = 9.25 and Standard Deviation = 5.93
7. Mean = 12.94 and Standard Deviation = 6.21
8. Mean = 14.45 and Standard Deviation = 5.78
9. Mean = 8.18 and Standard Deviation = 3.89
10. Mean = 14.8 and Standard Deviation = 6.31
11. Mean = 21.1 and Standard Deviation = 7.27
12. Mean = 17.47 and Standard Deviation = 5.24
13. Mean = 18.84 and Standard Deviation = 9.03
14. Mean = 20.6 and Standard Deviation = 9.95

Exercise 24D

1. (i) 11 (ii) 8.73 cm
2. (i) 14 (ii) £11.74
3. (i) 12 (ii) £1.92
4. (i) 4 (ii) 5.80 cm
5. (i) 25 (ii) 12.8 km
6. (i) 5 (ii) 4.23 kg
7. (i) 4 (ii) 9.83 m
8. (i) 3 (ii) 10.48 cm
9. (i) 6 (ii) 72.2 ml

Exercise 24E

1. Mean = 45.3 kg and Standard Deviation = 4.22 kg
2. Mean = 53.2 and Standard Deviation = 7.73
3. Mean = 43.1 and Standard Deviation = 7.32
4. Mean = £187.56 and Standard Deviation = £14.09
5. Mean = 5.2 cm and Standard Deviation = 0.395 cm
6. Mean = 65.6 and Standard Deviation = 6.576
7. Mean = 54.24 and Standard Deviation = 5.96
8. Mean = 50 cm and Standard Deviation = 7.29 cm

Exercise 24F

1. 16.2 cm
2. 12.5 kg
3. 7.75 cm
4. 9
5. 18.2 cm
6. £55.78
7. £5.09
8. 15.4 cm

9. 36.4 km
10. 11 kg
11. 26.7 cm
12. 109.6 ml

Exercise 24G

1. (i) 52 and 4.7 (ii) 92 and 9.4 (iii) 20 and 2.35
2. (i) 60.8 cm and 11.2 cm (ii) 10.2 cm and 2.8 cm
   (iii) 14.6 cm and 1.4 cm
3. (i) 13.5 and 0.725 (ii) 64 and 2.9 (iii) 268 and 14.5
4. (i) 0.9 mins and 0.24 mins (ii) 4.2 mins and 0.72 mins
   (iii) 0.95 mins and 0.12 mins
5. (i) 5.8 m and 1.2 m (ii) 2.6 m and 0.6 m
   (iii) 15.4 m and 3.6 m
6. (i) 5.9 cm and 2.31 cm (ii) 33.6 cm and 9.24 cm
   (iii) 4.2 cm and 0.77 cm
7. (i) 137 and 18 (ii) 38 and 6 (iii) 213 and 36
8. (i) 2.2 km and 0.3 km (ii) 5.6 km and 0.6 km
   (iii) 0.9 km and 0.075 km
9. (i) 7.43 and 0.27 (ii) 0.27 and 0.03 (iii) 3.29 and 0.81
10. (i) 2.6 l and 0.8 l (ii) 6.8 l and 1.6 l
   (iii) 0.83 l and 0.16 l

Exercise 25A

1. (a)

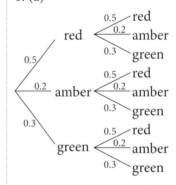

(b) (i) 0.3 (ii) 0.62 (iii) 0.51

2. (a)

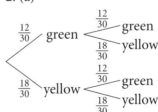

(b) (i) ⁶⁄₂₅ (ii) ⁴⁄₂₅ (iii) ¹²⁄₂₅

3. (a)

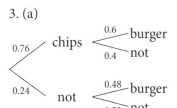

(b) (i) 0.456 (ii) 0.5712

4. (a)

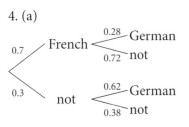

(b) (i) 0.196 (ii) 0.382

5. (a)

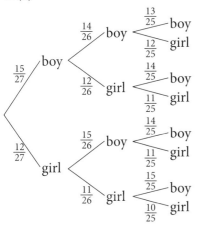

(b) (i) $^{22}/_{195}$ (ii) $^{541}/_{585}$ (iii) $^{242}/_{585}$

6. (a)

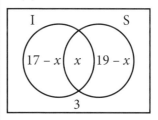

(b) (i) 0.1292 (ii) 0.6314

7. (a)

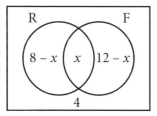

(b) (i) $^{18}/_{55}$ (ii) $^{28}/_{55}$

8. (a)

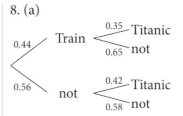

(b) (i) 0.154 (ii) 0.3892

Exercise 25B

1. (a)

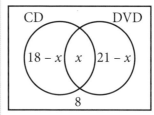

(b) 6 (c) (i) ⅓ (ii) ⅑

2. (a)

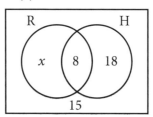

(b) 8 (c) (i) $^{9}/_{31}$ (ii) $^{12}/_{31}$

3. (a)

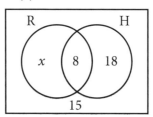

(b) 14 (c) (i) $^{8}/_{33}$ (ii) $^{14}/_{33}$

4. (a)

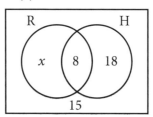

(b) (i) 14 (ii) 22 (c) (i) $^{26}/_{55}$ (ii) $^{3}/_{11}$

5. (a)

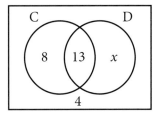

(b) 6 (c) (i) $^{13}/_{31}$ (ii) $^{6}/_{31}$

6. (a)

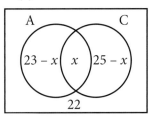

(b) 9 (c) (i) $^{14}/_{61}$ (ii) $^{9}/_{61}$

7. (a)

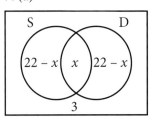

(b) 11 (c) (i) $^{22}/_{36}$ (ii) $^{11}/_{36}$

8. (a)

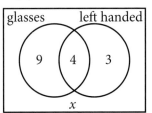

(b) 11 (c) (i) $^{4}/_{27}$ (ii) $^{1}/_{3}$ (iii) $^{1}/_{9}$

9. (a)

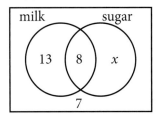

(b) 4 (c) (i) ¼ (ii) ⅛ (iii) $^{7}/_{32}$

10. (a)

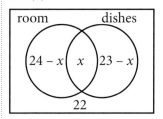

(b) 9 (c) (i) $^{3}/_{20}$ (ii) $^{7}/_{30}$ (iii) ¼

11. (a)

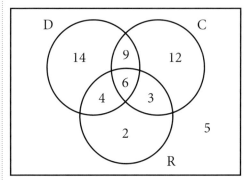

(b) (i) $^{6}/_{55}$ (ii) $^{5}/_{55}$ or $^{1}/_{11}$

12. (a)

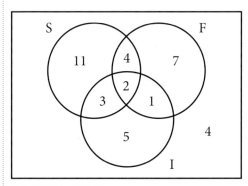

(b) (i) $^{14}/_{37}$ (ii) $^{1}/_{37}$

Exercise 25C

1. (a)

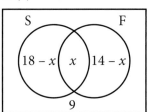

Study both = 8 (b) (i) $^{10}/_{33}$ (ii) $^{4}/_{7}$ (iii) $^{4}/_{9}$

180

2. (a)

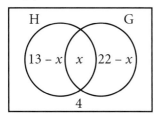

Study both = 5 (b) (i) ½ (ii) ⁵⁄₂₂ (iii) ⁵⁄₁₃

3. (a)

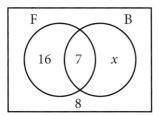

Black = 9 (b) (i) ²⁄₅ (ii) ⁷⁄₂₃ (iii) ⁷⁄₁₆

4. (a)

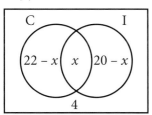

Order both = 7 (b) (i) ³⁵⁄₃₉ (ii) ⁷⁄₂₀ (iii) ⁷⁄₂₂

5. (i) ⁵⁄₆ (ii) ⁷⁄₁₂

6. (a)

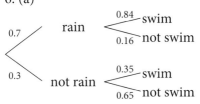

(b) (i) 0.588 (ii) 0.693 (iii) 0.84 (iv) 0.848

7. (a)

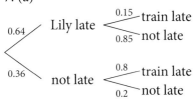

(b) (i) 0.096 (ii) 0.384 (iii) 0.25 (iv) 0.15

8. (a)

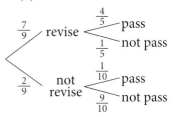

(b) (i) ²⁸⁄₄₅ (ii) ²⁹⁄₄₅ (iii) ⁴⁄₅

9. P(ice cream given that you get chips) = 0.15
P(chips given that you get ice cream) = 0.1

10. (a)

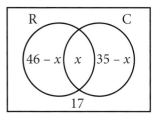

Went to both = 7 (b) (i) ¹⁄₁₃ (ii) ⅕
11. (i) 0.875 (ii) 0.625
12. (i) 0.778 (ii) 0.583
13. (i) ⁷⁄₁₁ (ii) ²⁰⁄₃₃

Exercise 25D

1. (i) 0.8 (ii) 0.5
2. (i) 0.22 (ii) 0.55
3. (i) 0.5 (ii) 0.8
4. (i) 0.5 (ii) 0.2
5. (i) 0.6 (ii) 0.525
6. (i) ⁸⁄₃₅ (ii) ⁴⁄₇
7. (i) 0.196 (ii) 0.35
8. (i) 0.0384 (ii) 0.16
9. (i) 0.3 (ii) 0.32
10. (i) 0.76 (ii) 0.6
11. (i) ⅜ (ii) ¹⁵⁄₅₆
12. (i) ³⁵⁄₇₂ (ii) ³⁵⁄₆₆
13. (i) ⁸⁄₁₃ (ii) ²⁄₇

Exercise 26A

1. (a) 0.607 (b) Positive
2. (a) 0.893 (b) Positive
3. (a) −0.557 (b) Negative
4. (a) 0.768 (b) Positive
5. (a) −0.583 (b) Negative
6. (a) −0.857 (b) Negative
7. (a) 0.571 (b) Positive
8. (a) −0.720 (b) Negative

9. (a) 0.720 (b) Positive
10. (a) –0.696 (b) Negative

Exercise 26B

Please note that in these answers the equations of the lines of best fit are approximate and depend upon the lines drawn by the student.

1. (a) (b)

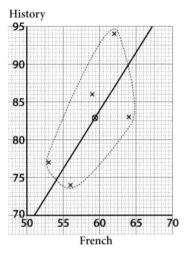

History / French

Mean (58.8, 82.8)
(c) $y = 1.64x – 13.64$
(d) 0.6
(e) Positive correlation

2. (a) (b)

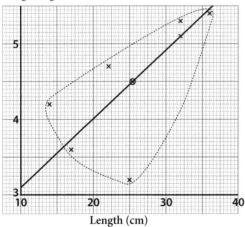

Weight (kg) / Length (cm)

Mean (25.4, 4.5)
(c) $y = 0.09x – 2.2$
(d) 0.74
(e) Positive correlation

3. (a) (b)

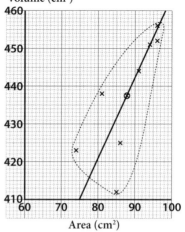

Volume (cm³) / Area (cm²)

Mean (87.875, 437.625)
(c) $y = 2.15x + 248.7$
(d) 0.875
(e) Positive correlation

4. (a) (b)

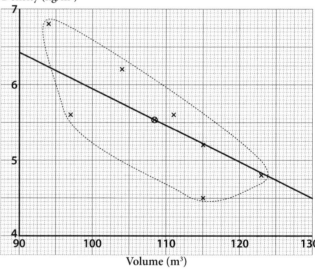

Density (kg/m³) / Volume (m³)

Mean (108.4, 5.53)
(c) $y = -0.047x + 10.62$
(d) –0.839
(e) Negative correlation

5. (a) (b)

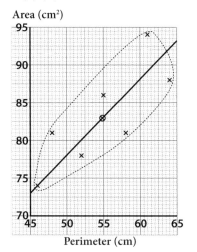

Mean (54.9, 83)
(c) $y = 1.01x + 27.6$
(d) 0.848
(e) Positive correlation

7. (a) (b)

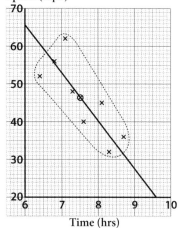

Mean (7.5375, 46.375)
(c) $y = -12.1x + 137.6$
(d) −0.857
(e) Negative correlation

6. (a) (b)

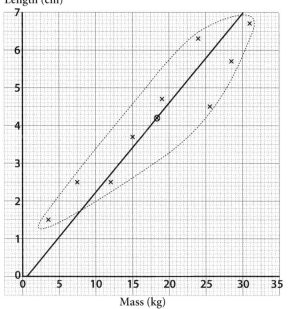

Mean (18.4, 4.2)
(c) $y = 0.239x - 0.2$
(d) 0.9125
(e) Positive correlation

8. (a) (b)

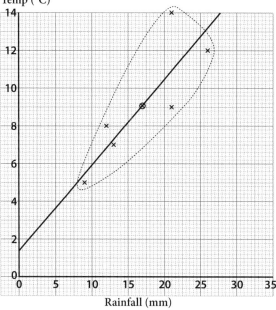

Mean (17, 9.17)
(c) $y = 0.46x + 1.4$
(d) 0.843
(e) Positive correlation

9. (a) (b)

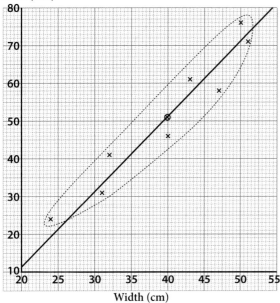

Mean (39.75, 51)
(c) $y = 1.97x - 27.4$
(d) 0.952
(e) Positive correlation

10. (a) (b)

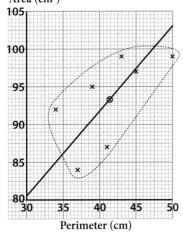

Mean (41.3, 93.3)
(c) $y = 1.13x + 46.6$
(d) 0.759
(e) Positive correlation